Data science is having profound effects on public and private life. However, thanks to a combination of technical complexity, proprietary privilege, and secretive state practice, both policy makers and the public remain largely in the dark about its uses and implications. This book offers desperately needed insights into the ways that Big Data analytics are being developed and applied in the domains of law enforcement, crime control and criminal justice. Alex Završnik has compiled a critically important collection of essays that shed light on the profound changes afoot in the way societies define, investigate, prosecute, and punish crime and criminality. These scholars are sounding the alarm about the major challenges that Big Data poses to civil rights and social justice, unpacking some of the ways that new systems of data analytics are undermining legal concepts and tenets that are fundamental to democratic governance. The volume covers considerable ground: how predictive policing and algorithmic sentencing exacerbate racial and class-based discrimination; how data analytics not only fails to prevent but enables financial crimes and tax evasion among economic elites; how "informed consent" morphs into "forced consent" with ubiquitous data tracking; how automated policing pushes toward total information capture; how different legal concepts of privacy have shaped the possibilities of judicial oversight of mass data collection; how international law addresses cyber-espionage; and more. It is a must-read for anyone concerned with how Big Data and predictive analytics are disrupting and destabilizing the institutions and ideals of democracy.

Kelly Gates, *Department of Communication and Science Studies Program, University of California San Diego, USA*

Big Data, Crime and Social Control

From predictive policing to self-surveillance to private security, the potential uses of big data in crime control pose serious legal and ethical challenges relating to privacy, discrimination and the presumption of innocence. The book is about the impacts of the use of big data analytics on social and crime control and on fundamental liberties.

Drawing on research from Europe and the US, this book identifies the various ways in which law and ethics intersect with the application of big data in social and crime control, considers potential challenges to human rights and democracy and recommends regulatory solutions and best practice. This book focuses on changes in knowledge production and the manifold sites of contemporary surveillance, ranging from self-surveillance to corporate and state surveillance. It tackles the implications of big data and predictive algorithmic analytics for social justice, social equality and social power: concepts at the very core of crime and social control.

This book will be of interest to scholars and students of criminology, sociology, politics and socio-legal studies.

Aleš Završnik is a Senior Research Fellow at the Institute of Criminology at the Faculty of Law and Associate Professor at the Faculty of Law, University of Ljubljana, Slovenia.

Routledge Frontiers of Criminal Justice

www.routledge.com/Routledge-Frontiers-of-Criminal-Justice/book-series/RFCJ

43 Young Offenders and Open Custody
Tove Pettersson

44 Restorative Responses to Sexual Violence
Legal, Social and Therapeutic Dimensions
Edited by Estelle Zinsstag and Marie Keenan

45 Policing Hate Crime
Understanding Communities and Prejudice
Gail Mason, JaneMaree Maher, Jude McCulloch, Sharon Pickering, Rebecca Wickes and Carolyn McKay

46 The Special Constabulary
Historical Context, International Comparisons and Contemporary Themes
Edited by Karen Bullock and Andrew Millie

47 Action Research in Criminal Justice
Restorative Justice Approaches in Intercultural Settings
Edited by Inge Vanfraechem and Ivo Aertsen

48 Restoring Justice and Security in Intercultural Europe
Edited by Brunilda Pali and Ivo Aertsen

49 Monitoring Penal Policy in Europe
Edited by Gaëtan Cliquennois and Hugues de Suremain

50 Big Data, Crime and Social Control
Edited by Aleš Završnik

51 Moral Issues in Intelligence-led Policing
Edited by Nicholas R. Fyfe, Helene O. I. Gundhus and Kira Vrist Rønn

52 The Enforcement of Offender Supervision in Europe
Understanding Breach Processes
Edited by Miranda M. Boone and Niamh Maguire

53 Diversion in Youth Justice
What Can We Learn from Historical and Contemporary Practices?
Roger Smith

Big Data, Crime and Social Control

Edited by Aleš Završnik

Routledge
Taylor & Francis Group
LONDON AND NEW YORK

First published 2018 by Routledge

2 Park Square, Milton Park, Abingdon, Oxfordshire OX14 4RN
52 Vanderbilt Avenue, New York, NY 10017

Routledge is an imprint of the Taylor & Francis Group, an informa business

First issued in paperback 2019

British Library Cataloguing in Publication Data
A catalogue record for this book is available from the British Library

Library of Congress Cataloging in Publication Data
A catalog record for this book has been requested

ISBN: 978-1-138-22745-3 (hbk)
ISBN: 978-0-367-22756-2 (pbk)

Typeset in Times New Roman
by Wearset Ltd, Boldon, Tyne and Wear

Contents

Contributors

Editor

Aleš Završnik, Doctor of Law (LLD.), is a Senior Research Fellow at the Institute of Criminology at the Faculty of Law in Ljubljana and Associate Professor at the Faculty of Law, University of Ljubljana. He was a postdoctoral fellow at the University of Oslo (2012) and at the Max-Planck-Institute für ausländisches und internationals Strafrecht, Freiburg i. Br. (2009), and is starting a Visiting Fellowship at the Collegium Helveticum Zürich, a joint initiative of the ETH Zürich and the University of Zürich (2017–18). He collaborated in several European Cooperation in Science and Technology (COST) Actions, e.g. *Living in Surveillance Societies*. In his latest research Završnik focused on surveillance implications of drones in the book he edited, *Drones and Unmanned Aerial Systems: Legal and Social Implications for Security and Surveillance*. He also co-edited a book, *Crime and Transition in Central and Eastern Europe* that was awarded the best scientific achievement in criminology by the Slovenian Research Agency in 2012. He has extensively researched and published on cybercrime, IT-law, surveillance, crime control and technology. Završnik conducts ethical analysis for security and ICT projects, e.g. he is an independent Ethics Expert with REA, the research arm of the European Commission, for Horizon 2020 projects. Among others, he led a research project *Law in the Age of Big Data: Regulating Privacy, Transparency, Secrecy and Other Competing Values in the 21st Century* (funded by the Slovenian Research Agency, No. J5–6823). Email: ales.zavrsnik@pf.uni-lj.si.

Contributors

Mark Andrejevic, Associate Professor of Media Studies at Pomona College, Claremont, California, USA. He is a media scholar who writes about surveillance, new media and popular culture. In broad terms, he is interested in the ways in which forms of surveillance and monitoring enabled by the development of new media technologies impact the realms of economics, politics and culture. His first book, *Reality TV: The Work of Being Watched* (2003), explores the way in which this popular programming genre equates participation with willing submission to comprehensive monitoring. His second book, *iSpy: Surveillance and Power in the Interactive Era* (2007), considers the role of surveillance in the era of networked digital technology and explores the consequences for politics, policing, popular culture, and commerce. His third book, *Infoglut: How Too Much Information Is Changing the Way We Think and Know*, explores the social, cultural, and theoretical implications of data mining and predictive analytics. His work has appeared in edited collections and in academic journals including *Television and New Media; New Media and Society; Critical Studies in Media Communication; Theory, Culture & Society; Surveillance & Society; The International Journal of Communication; Cultural Studies; The Communication Review*, and the *Canadian Journal of Communication.* His current work explores the logic of automated surveillance, sensing, and response associated with drones.

Primož Gorkič. After attaining a BA in law in September 2002, Primož Gorkič enrolled in postgraduate studies in Civil and Commercial Law at the Ljubljana Faculty of Law, later substituted for studies in Criminal Law. He attained his doctoral degree in November 2009. Finishing his traineeship with the Ljubljana Higher Court in November 2004, he applied for the position of assistant–researcher at the Ljubljana faculty. He passed the national bar exam in April 2005. From October 2007 till November 2009 he worked as part-time advisor to the Supreme Court of Slovenia. He is currently working as Associate Professor with the Chair for Criminal Law at Ljubljana Faculty of Law and part-time researcher at the Institute of Criminology at Ljubljana Faculty of Law. His research focuses on criminal procedure, evidence law and human rights in the context of criminal justice. He teaches criminal procedure law (undergraduate and graduate studies) and criminal investigation (graduate studies). He currently presides over the Slovenian Association for Criminal Law and Criminology.

Zoran Kanduč, Doctor of Law (LLD), Associate Professor, is a Research Counsellor at the Institute of Criminology at the Faculty of Law Ljubljana. He lectures on criminology and victimology at the Faculty of Criminal Justice and Security (University of Maribor) and at the Faculty of Law (University of Ljubljana). He has extensively researched and published on criminology theories, crime, social and class control in modern and post-modern societies, crime in post-socialist social formation, and various forms of "structural" and personal violence. His monographs (in the Slovene language) are "Criminology: Straight and Deviant Ways of Science of Straight and Deviant Ways" (1999), "Beyond Crime and Punishment" (2003), "Subjects and Objects of Social Control in the Context of Post-Modern Transitions" (2007) and "Politics, Law, Economy and Crime: Criminological Reflections of Post-Modern Society and Culture" (2013). He is a member of the editorial board of two Slovene scientific reviews covering criminology, victimology and security studies.

Maša Kovič Dine is a PhD. candidate and a Teaching Assistant at the Department of International Law, Faculty of Law, University of Ljubljana. After receiving her graduate law degree in Slovenia, she obtained her LLM. degree from the Faculty of Law at the University of Toronto and worked at the University of Toronto G8 Research Group, analysing G8 countries' compliance with their summit commitments. Her research interests include international environmental law, cybersecurity and responsibility to protect. She is a member of the International Law Association and regularly acts as a Slovenian national expert in various international research projects.

Frank Pasquale is Professor of Law at the University of Maryland, Carey School of Law, an Affiliate Fellow at Yale Law School's Information Society Project, and a member of the Council for Big Data, Ethics and Society. Pasquale has been a Visiting Fellow at Princeton's Center for Information Technology, and a Visiting Professor at Yale Law School and Cardozo Law School. He was a Marshall Scholar at Oxford University. Frank Pasquale's research addresses the challenges posed to information law by rapidly changing technology, particularly in the health care, internet and finance industries. He frequently presents on the ethical, legal, and social implications of information technology for attorneys, physicians, and other health professionals. His book *The Black Box Society: The Secret Algorithms that Control Money and Information* (Harvard University Press, 2015) develops a social theory of reputation, search, and finance.

Mojca M. Plesničar, Assistant Professor and Research Fellow at the Institute of Criminology at the Faculty of Law in Ljubljana. She concluded her postgraduate studies at the Centre for Criminology University of Oxford and her PhD in Criminology at the Faculty of Law Ljubljana. Her main areas of research are sentencing and punishment, but she also deals with questions of technology and development in connection to criminal justice. She teaches criminology, penology and English legal terminology at the University of Ljubljana and University of Maribor. She regularly presents papers at international scholarly gatherings and has published several articles and chapters in international publications on sentencing and related subject, most recently she has co-authored a chapter on "Sentencing, Legitimacy, and Public Opinion" in the collective book *Trust and Legitimacy in Criminal Justice: European Perspectives* (Springer, 2015) and written the article "Good and Not so Good Reasons for Differences in Sentencing" (*Revija za kriminalistiko in kriminologijo*, 2015).

Renata Salecl is a Slovene philosopher, sociologist and legal theorist. She is a senior researcher at the Institute of Criminology, Faculty of Law at the University of Ljubljana, and holds a professorship at Birkbeck College, University of London. She has been a visiting professor at the London School of Economics, lecturing on the topic of emotions and law. Every year she lectures at Benjamin N. Cardozo School of Law (New York), on Psychoanalysis and Law, and she has also been teaching courses on neuroscience and law. From 2012, furthermore, she has been visiting professor at the Department of Social Science, Health and Medicine at King's College London. Her books include: *Sexuation*. Durham, North Carolina: Duke University Press, 2000; *On Anxiety*. London New York: Routledge, 2004; and *Choice*. London: Profile, 2010. The books have been translated into 13 languages.

Katja Šugman Stubbs holds degrees in Law and Psychology and a PhD in Law. She is Professor of Criminal Law and Criminology at the Faculty of Law, University of Ljubljana, Slovenia where she teaches criminal procedure, EU criminal law, evidence law, criminology, and law and psychology. She is a Slovenian member of the Council of Europe Committee for the prevention of torture and a member of the Slovenian Prosecutorial Council. Her main research interests lie in the field of human rights in criminal procedure, balance between freedom and security in society, and analysis of the psychological factors which influence judicial decisions. She was head of the research team drafting a new model of the Slovenian procedure code and collaborated in numerous national and international research projects (e.g. The future of mutual recognition in criminal matters in the European Union [ECLAN, Université libre de Bruxelles, Belgium], preparatory study for an impact assessment on a new legislative instrument replacing Council Framework Decision 2004/757/JHA on illicit drug trafficking). She has published books and articles on a broad range of topics from analysis of different aspects of the rights of the accused, exclusion of evidence, assessment of EU crime policies, the European Arrest Warrant to criminological topics. She was a Visiting Professor at Cambridge (Fitzwilliam College, UK) and at the Institute de sciences criminelles, Université de Poitiers (France).

Maruša T. Veber is a Teaching Assistant and a PhD candidate at the Department of International Law at the Faculty of Law, University of Ljubljana. Previously she was a junior visiting fellow at the Graduate Institute of International and Development Studies in Geneva, a visiting scholar at the University of Hull Law School and a researcher at the United Nations University Institute on Comparative Regional Integration Studies (UNU-CRIS) in Bruges. She is a member of the Manchester International Law Centre, the Geneva International Sanctions Network and the Slovene branch of the International Law Association. She is currently part of the research project *Law in the Age of Big Data: Regulating Privacy, Transparency, Secrecy and Other Competing Values in the 21st Century* (Slovenian Research Agency, 2014–17), led by the Institute of Criminology at the Faculty of Law in Ljubljana, where she focuses on the international law aspects of cyber espionage between states.

Dean Wilson, Professor, completed his undergraduate and initial postgraduate studies at the University of Auckland, New Zealand. His Masters thesis was a historical study of violence, the law and local community in a colonial settlement. Dean continued postgraduate study in Australia, gaining a scholarship to Monash University where he undertook the first major historical study of urban policing in Australia. Having completed his PhD in 2001, Dean has subsequently concentrated on contemporary criminological research, including research into surveillance, technology and security, policing and victims of crime and border militarization and policing. Dean's research and teaching engage interdisciplinary methodologies that traverse the fields of criminology, sociology, media, law, history and politics. Dean was a Lecturer and then Senior Lecturer in Criminology at Monash University, Melbourne, Australia between 2003 and 2010, and a Reader in Criminology at the University of Plymouth prior to joining the Department of Sociology at the University of Sussex in 2015.

Foreword

Automating criminal justice

Katja Franko
University of Oslo

Keeping up with the fast pace of surveillance and data collection is one of the main political challenges of our time. We are dealing here with a phenomenon that is so massive, that changes so quickly, and has become so ubiquitous, that it seems impossible to pin it down, not least through legal regulation. The law seems to run against the grain of social change and massive political pressures. This book, however, is an attempt to do just that. It is an impressive piece of work and exemplary in its pursuit of a clear and specific issue – surveillance and big data – through many thematic perspectives that complement each other. While addressing a topic of great public importance, the analysis is conscientious and meticulous, yet ambitious and at times innovative. The book contains a rich set of ideas and concepts, and is well referenced to a substantial and diverse literature. The authors make good use of important conceptual devices to help us think more deeply about the problem of big data and automation of criminal justice and human life more generally.

The editors are to be applauded for including a rich array of diverse topics, from automated and predictive policing, big data industry, artificial intelligence, and many more. By drawing on such diverse sets of examples the book produces a much more complete view of big data than the usual particularistic studies that dominate the literature. In this way, the book disaggregates the seemingly plain initial dilemma between surveillance and privacy, and adds layers of complexity and contextualization. The book raises a number of interesting questions, and the analysis identifies a series of innovative tensions between, among other, free will and modern society, the value of privacy versus urban safety, the objectives of law enforcement versus individual rights. The rapidly changing relationships between humans and machines are radically transforming the nature of contemporary criminal justice.

It is worth taking note of the range of literature that is included, from legal and socio-legal perspectives to classical surveillance studies, urban studies and philosophy. However, unlike several other books in the field, this diversity does not come at a cost in terms of cohesion. From a scholarly point of view, the work will be valuable for legal and social-legal scholars, sociologists and other social scientists, as well as students of political science and others generally interested in issues of surveillance.

Acknowledgements

I wish to express my gratitude to my co-authors and to the many people who have made this book possible.

This book is intended to be a timely contribution to the burgeoning field encompassing the legal, ethical and social aspects of big data, artificial intelligence, robotics and automation. Delving into the intersection of these fields would have been much less exciting without the contributors to this book. I am grateful for the exchange of ideas and the advice to take a step backwards rather than two steps forward in order to grasp the "bigger picture" regarding our increasingly automated and de-subjectivised form of digital surveillance capitalism. Without exception, the contributors have demonstrated great sensitivity as to social justice and have detected profound changes in the power/knowledge relations provoked by big data tools and algorithmic-driven decision-making protocols. Throughout this project, I have felt that we share the belief that the algorithmic turn has not enabled bigger and better things to happen, but has done more to isolate, analyse and discriminate against individuals and deepen the existing divisions in our societies. Several contributors to the book are, if I may say so, superstars in their field of research and it has thus been an honour to work alongside them and a pleasure to observe the social sensitivity and erudite thoughts they shared while mapping and framing the world in the light of this intangible and elusive concept called "big data". I have great admiration for their work. I worked with some of them for the first time and they were courageous enough to trust the editing of the book to me. With others, I was thankful to be able to work with them again, as I admire and appreciate their work and thoughts.

My gratitude also goes to my colleagues at the Institute of Criminology at the Faculty of Law of the University of Ljubljana. They have often been sceptical of the new "algorithmic society" and "algocracy", which motivated me to sharpen my arguments, but they were fully supportive of the book project throughout, for which I am grateful. My special gratitude goes to Professor Renata Salecl, a contributor to the book and the leader of the research programme at the Institute of Criminology. She has built amazing bridges between continental and Anglo-American scholars, as well as along many other north–south and east–west vectors, from which I have greatly benefited. My thanks also go to Professor

Matjaž Jager, Director of the Institute of Criminology, for his full support on this book project.

I would like to express my gratitude to the many researchers whom I met with over the last two years but cannot name individually; I appreciate their remarks, observations and inspiration provided at conferences, consultations and other events. In particular, I would like to thank the leadership of the European Group for the Study of Deviance and Social Control, an international network working towards, inter alia, social justice and state accountability, which dedicated a special stream to the surveillance future at the 43rd annual conference in Tallinn in 2015, where I met some of the book contributors for the first time and held initial discussions with the publisher's editor.

I am very grateful to Professor Jure Leskovec from the University of Stanford for the excellent colloquium we held in Ljubljana in 2015 dedicated to a simple provocative question "Can a computer judge better than a human judge?" He was willing to openly share his cutting-edge insights into the big data analytics deployed in criminal justice settings and raised many questions and helped us frame further issues related to algorithms and statistical modelling in criminal justice.

I would like to express my gratitude to the many interlocutors at the Privacy and Surveillance Sections at the Cyberspace Conference organised annually by the Institute of Law and Technology, the Faculty of Law and the Faculty of Social Studies of Masaryk University in Brno, Czech Republic, and its co-founder Dr Radim Polčák.

My gratitude also goes to Professor Zoran Kanduč for initiating and organising a bi-annual national Slovenian criminological conference devoted to the topic of the book in 2015. I would like to express my great appreciation to the Slovenian Association for Legal and Social Philosophy and the Department for Legal Theory and Sociology of Law at the Faculty of Law of the University of Ljubljana, which enabled me to share my research insights into algorithmic power at their annual Symposium on Legal and Social Philosophy in 2015, and especially to Professor Marijan Pavčnik, the President of the Association, and Professor Tilen Štajnpihler, the Secretary of the Department.

As regards outreach activities, I am grateful to Lenart Kučić, one of the most prominent journalists addressing the digital society in Slovenia, who works for the Delo daily newspaper; he provided invaluable support in disseminating various ideas and raising awareness of the role algorithms play in our society. Special thanks go to the Information Commissioner of the Republic of Slovenia, Mojca Prelesnik, and the Deputy Information Commissioner, Andrej Tomšič, MA, for providing timely updates from the battlefield for privacy, and for a special event held in the Slovenian Parliament to mark European data protection efforts in January 2016. I would also like to express my gratitude to the President of the Republic of Slovenia for being invited to consult on "Slovenia 2030", which is specifically dedicated to opportunities for digital transformation.

My research talks have spread nationally with the help of GV Publishing House and Info House – Institute for Privacy and Access to Public Information, e.g. the results were presented at the 9th Conference on Criminal Law and Criminology and at the 3rd Conference on Privacy and Freedom of Expression in 2017. Special thanks go to Dr Maja Bogataj Jančič, LLM., LLM., and Dr Matej Kovačič for sharing the podium a number of times at various conferences.

These events, contacts and discussions shaped my views and encouraged me to deepen my understanding of the ramifications of big data analytics and algorithmic power for the criminal justice system, policing and society at large.

Great efforts were made by the anonymous reviewers of the early draft of the book, who demonstrated profound understanding of the critical issues of the algorithmic turn and provided constructive suggestions.

The Slovenian Research Agency financially supported the research project "Law in the age of big data: Regulating privacy, transparency, secrecy and other competing values in the 21st century" (No. J5–6823), which enabled several of the authors to contribute to this book.

And finally, special appreciation goes to Dean J. DeVos, our English language editor, for his patience and diligence during our collaboration, and to Lara Brecelj, Klara Cvar, Neja Domnik and Maša Gril, who assisted with the research in the final stages of the preparation of the book. I would like to specifically thank the editorial team at the publisher, Hannah Catterall and Thomas Sutton, for being very supportive and encouraging throughout the whole process of preparing the book.

Aleš Završnik, Editor
Ljubljana, 14 April 2017

Part I

Introduction

1 Big data

What is it and why does it matter for crime and social control?

Aleš Završnik

> Big data is a cliché, because people think it is this magic mountain of gold. It's not a mountain of gold. Most of it is trash.
>
> Todd Yellin, Netflix (Yellin, 2016)

Big data semantics: "meaning extraction"

If the limits of our language entail the limits of our worlds (Wittgenstein, 2005), the language of big data is tearing down the world of what counts as crime-relevant knowledge (now databases), what counts as proper reasoning (now algorithms) and how we should tackle – prevent and investigate – crime (now predictive policing) and prosecute cases (now automated justice). The new mathematical language serves security purposes well (Amoore, 2014: 426). "[T]he mathematical sciences offer a grammar of combinatorial possibilities that allows for things – people, objects and data – to be arranged together, for links to be made." (Amoore, 2014: 431) New concepts and tools are being invented in order to understand crime (knowledge production) and act upon this knowledge (crime control policy): "meaning extraction", "sentiment analysis", "opinion mining", "computational treatment of subjectivity in text", visualisation and mining tools, algorithmic prediction, and data analytics. All these are rearranging and blurring the boundaries in the security and crime control domain.

The challenges of big data are to accumulate large quantities of data and also to clean and structure the data in order to extract meaningful actionable instructions – but for whom and at what price? These are never completely objective, value-free data that can speak for themselves. Already at the language level, researchers have shown how natural language necessarily contains human biases. The training of machines, also known as machine learning, on language corpora means that artificial intelligence (AI) will inevitably imbibe these biases as well (Caliskan-Islam, Bryson, & Narayanan, 2016). The researchers claim that the process of "de-biasing" cannot help eliminate biases either. The result would only be "fairness through blindness" because "prejudice can creep back in through proxies" (Caliskan-Islam et al., 2016). Research on sentencing prediction instruments has confirmed exactly how criminal history is in fact a proxy for race (Harcourt, 2015): the racial disproportionality in the prison population hits the African-American community hardest, in large part due to sentencing instruments being based on analysis of past criminality.

The new language and semantic shifts in crime control threaten fundamental civil liberties. "The start and finish of the criminal justice process are now indefinite and indistinct" (Marks, Bowling, & Keenan, 2015), and this extends – in depth and breadth – the scope of the surveillant gaze (Marx, 2002). Agencies mandated to fight terrorism, for instance, are now in pursuit of spotting "persons of interest", meaning not only what a person might be, but what a person might become (Lyon, 2014: 7). Similarly, the Eindhoven Living Lab in the Netherlands checks for "escalated behaviour" to arrive at big-data-generated determinations of when to intervene (Galič, 2017). The notion includes yelling, verbal or otherwise aggressive behaviour, or showing signs that a person has lost self-control (de Kort, 2014). Such behaviour is targeted and mediated in order to be "diffused". The goals of the intervention are to induce subtle changes, e.g. "a decrease in the excitation level", to refocus attention, to encourage social behaviour, or to increase the self-awareness and self-control of an individual or group (de Kort, 2014). New means and techniques are used to "de-escalate" behaviour, for instance, interactive street lighting manipulation is used "to confine and contain" aggressive events (de Kort, 2014).

These examples show how big data is used as a means of collecting and processing environmental data in urban settings, which are perceived as living organisms. The "living lab" automatically collects and analyses data, and in the process learns from past experience in order to produce actionable data. Big data capabilities are not centred on notions of solidarity, the social cohesion of the community, or other public interests, but are instead grounded in the monetary gains of homeowners and employed to guard the economic interests of retailers. The central goals of BTC Ljubljana's living lab are to increase the spending of customers, increase the profits of retailers, and make management more effective (Polajžer, 2016). Big data then supports the specific interests of certain privileged social groups. It is employed to seduce consumers into fulfilling the monetary ends of retailers, while public safety considerations are a mere side effect of these efforts.

Instead of the well-defined concepts in criminal law, such as suspect, reasonable doubt, the presumption of innocence, etc., which serve as regulators of and thresholds for the intervention of law enforcement agencies, the new concepts and language no longer sufficiently confine agencies nor prevent abuses of power. The language of big data helps to tear down the walls of criminal procedure rules. This move towards a system of "automatic justice" "minimise[s] human agency and undercuts the due process safeguards built into the traditional criminal justice model". (Marks et al., 2015).

Big data knowledge: "We do not know what the questions were, but here are the answers"

The meaning of "big data" has been highly contested and used to encompass the volume of datasets, the processing of data ("data mining", "crunching numbers") or even the generalised "big data hype" (Raicu, 2015). The most far-fetched definitions contend that big data analytics enables an entirely new epistemological approach to making sense of the world (Kitchin, 2014: 2). More modest authors claim that the idea of big data is "to see hidden connections and patterns" and generate new knowledge about existent data (The Information and Privacy Commissioner of Ontario, 2017). Big data "refers to our burgeoning ability to crunch vast collections of information, analyse it instantly, and draw sometimes profoundly surprising conclusions from it" (Mayer-Schönberger & Cukier, 2013).

The more far-fetched views see big data and the new analytical tools as a harbinger of a new paradigm shift by which science is entering into "the fourth stage" (Hey, Tansley, & Tolle, 2009). According to such views, big data signals a new era of knowledge production characterised by "the end of theory": "the data deluge makes the scientific method obsolete" (C. Anderson, 2008). The big data enterprise claims that rather than testing a theory by analysing data, the new analytics seeks to gain insight "born from the data." This "empiricist epistemology" (Kitchin, 2014) is based on several premises (Kitchin, 2014: 4–5): (1) the premise that big data captures the whole domain it seeks to analyse; (2) that there is no need for a priori theory; (3) that "the data speak for themselves" (Captain, 2015) free of human bias or framing; and (4) that the calculated meaning transcends context or domain-specific knowledge and can be interpreted by anyone who can decode visualisations. Such views camouflage big data as an "objective" and "pure" knowledge, and neglects the fact that statistics have always been political and served specific political ends (Desrosières, 2002). Statistics are produced by humans and for humans.

On the other hand, authors have warned that we are witnessing more of the old rather than something completely new. Peter Thiel, the founder of PayPal, has, paradoxically, pointed to the failures of the digital industry: "Cell phones distract us from the fact that the subways are 100 years old" (Dowd, 2017). The digital turn has failed to fulfil the old dreams that bigger and better things would happen. What may be new is the exponential acceleration of neoliberalism: a reduction in state power that reinforces the private sector; an increase in social and wealth inequalities by enabling the powerful elite to gain insight into the masses more than ever before; the expansion of the precarious working class, and an increase in unpaid digital labour. "The people" as the holders of sovereign rights, are reduced to "users" of digital services. They are blinded by the informational glut (Andrejevic, 2013), addicted to the comforting and entertaining nature of digital services in order not to see the real interests behind the user-friendly interfaces of digital applications and the high stakes that are (not) being negotiated. This army of digital workers open to exploitation is used as the means to very specific political ends. There is no "end of politics" at work here, as the "reserve army of digital labour" serves the pecuniary interests of the digital industry, which caters to the affluent elite of the surveillance society. Or, to paraphrase a popular meme, digital workers are actually the product being sold on the data marketplace.

Big data industry: "Doing more with less"

Metaphors are important for technological progress to have an impact on society (Watson, 2015). Big data's origins can be traced to the domain of business and industry, where big data has been defined as three "V"s (Gartner, 2012), four "V"s (IBM, 2016), or even five or six "V"s (Marr, 2016), regarding its: (1) volume (the scale of the data gathered is greater); (2) velocity (data processing and analysis of streaming data takes place in real time or near real time; processing is faster and enables longer retention and faster transfer of data); (3) the variety of data (in terms of different forms of data, structured and unstructured data, computer readable or not, e.g. social medial posts); (4) veracity (the uncertainty of data, which refers to the quality and accuracy of data [Ramesh, 2017]); (5) the "value" of data; and (6) its "vulnerability" (Marr, 2016).

In order to market big data tools, the industry has depicted data through a variety of words and metaphors (Watson, 2015): data can be thought of in terms of (1) a new natural resource (in the same vein as oil, gold, etc.); (2) a new industrial product (as big business, a platform, etc.); (3) a new by-product (e.g. data trails, data smog, data exhaust, etc.); (4) a new market (with its currencies, brokers, vaults, assets, etc.); (5) a liquid (thus: data deluges, data tsunamis, data waves, data lakes); (6) being trendy (the new oil, the new currency, the new black, the new revolution, etc.); and (7) a body (through reference to words such as fingerprint, blood, DNA, reflection, shadow, profile, and portrait).

The application of these vague concepts obfuscates the fact that big data is designed to serve certain political and economic interests over others, e.g. to increase the profits of companies by finding "hidden opportunities" in production cycles, or in government operations – to put it simply with one of the sale pitches, "to do more with less".

The industrial origin of big data carries the logic of business practices into all other sectors that apply big data. This results in different deficiencies and damage in these domains that may be acceptable in one domain (e.g. marketing), but which erode rights and deny human dignity in another. For instance, when employers increasingly indulge in the practice of "hiring by algorithm", seemingly irrelevant data such as inaccurate consumer reports can cause real economic injury to job seekers. Such inaccuracies can lead employers to screen out prospective employees or to lower salaries (Cohen, Hoofnagle, McGeveran, Ohm, & Reidenberg et al., 2015). In the crime control domain, such practices breach several fundamental principles of criminal procedure (Mosco, 2014: 177). The pressure to institute a process of "datafication", i.e. turning everything into numbers, in order to "monetise" data and create "actionable" data, is at the core of big data logics. The implicit underpinnings of business imperatives in the security domain have been described as three "A"s – automation, anticipation, and adaptation (Lyon, 2014: 6–9).

Anticipation refocuses crime control actors. They reorient their practices and:

> focus on the future more than on the present and the past. In the context of neo-liberal governance, this anticipation is likely to place more weight on

> surveillance for managing consequences rather than research on understanding causes of social problems such as crime and disorder.
>
> (Lyon, 2014: 6–8)

Big data fortifies "evidence-based" policing, a correlate of which can be found in penology as "evidence-based" sentencing or "truth in sentencing". The common feature of these changes is a higher reliance on and increased trust given to quantitative over more qualitative reasoning. In fact, prediction – predictive analytics that transcend human perception (Siegel, 2013) – has been one of the most attractive aspects regarding the application of big data in crime control. It is powered by data, which is being accumulated in large part as the by-product of routine tasks (Siegel, 2013). In addition to their effects on criminal procedure, such practices involving bulk data collection from different sources – even before determining the full range of their actual and potential uses (Lyon, 2014: 4) – breach the fundamental principles of personal data protection (i.e. the proportionality principle, the principle of purpose specification and minimisation, and obtaining valid prior consent from data subjects).

Algorithms are being used "not only to understand a past sequence of events, but also to predict and intervene before behaviours, events, and processes are set in train (*sic*) [i.e. motion]" (Lyon, 2014: 4). Prediction is only the first step and is followed by pre-emption – taking action in order to prevent an anticipated event from happening. This is causing existing *ex post facto* criminal policy to adopt *ex ante* preventive measures (Kerr & Earle, 2013). It may also invoke "a feeling of constant surveillance", as stressed by the European Court of Justice in the case *Digital Rights Ireland Ltd* v. *Ireland* (C-293/12 and C-594/12, dated 8 April 2014) on the bulk collection of the traffic and location data of users of all types of electronic communications.

Automated governance

In the much acclaimed book *Sentencing in the Age of Information*, Franko Aas (2005) detected how biometric technology used in criminal justice settings had been changing the focus from "narrative" to supposedly more "objective" and unbiased "computer databases". Franko was critical of the transition and characterised it as going "from narrative to database". Instead of the narrative, she claimed, the body has come to be regarded as a source of unprecedented accuracy because "the body does not lie" (Aas, 2006). The "truth" will supposedly be detected on a body through the use of biometric technologies, such as DNA and fingerprinting, because "coded bodies" had come to be perceived as a more reliable means of ascertaining the truth, in comparison to witness testimonies, which research has shown how misleading they can be.

However, today criminal justice systems are on the brink of taking yet another step: from the database towards automated algorithmic-based decision-making. This is a transition towards complete de-subjectivation in the decision-making process, a sort of erasure of subjectivity. It is not only that narrative is regarded as an unreliable means to discover the "truth in sentencing", as many have claimed from different perspectives, e.g. from psychological and neurological points of view (Bradfield, Wells, & Olson, 2002; Innocence Project, 2016; Kassin, 2008; Kassin & Kiechel, 1996; Shaw & Porter, 2015), but even databases are regarded as being too weak and static, too dependent on the user. Big data promises to make this static tool "actionable" through use of algorithms that can provide real-time feedback by crunching large amounts of data coming from all domains of life. At the same time as information collection is expanding, data processing is getting faster; responses are becoming (semi-)automated, subjectivity is being expelled, and the real power is being transferred from the democratic polis to the digital corporation.

The substitution of human judgement with AI tools is then part of a more fundamental shift in the art of governing society. The power of Cambridge Analytica, a data company employed by Donald Trump in the 2016 American presidential elections, offers insight into the possible political power algorithms might acquire in the future: "a Weaponized AI Propaganda Machine [...] used to manipulate our opinions and behaviour to advance specific political agendas. [...] [An] invisible machine that preys on the personalities of individual voters to create large shifts in public opinion" (Anderson & Horvath, 2017). The company gathered 5,000 data points on every person in the US and used this to psychologically profile people, and deliver highly personalised advertising online. This exploited voters' characters, fears, and interests. And this swung the election to Trump. Despite convincing criticism that the company could not have had such an amount of detailed personal data (Blackie, 2017), this may be the future of algorithmic governance and politics. However, what has been made clear is how the shift in data collection, analysis, and knowledge about society from the public sphere to the private sector can affect governance at the national level and corrupt the democratic architecture of a state.

This can be unambiguously observed in an existing "living social laboratory", Singapore, where big data fantasies have been most forcefully brought to life (Harris, 2014). The so-called "total information awareness systems" (TIAS) started as a pandemic prediction tool after the SARS outbreak which occurred from late February to July of 2003, and over the years the system has grown into a tool for "canvassing a range of sources for weak signals of potential future shocks" (Harris, 2014). TIAS uses a mixture of proprietary and commercial technology based on a "cognitive model" designed to mimic the human thought process (Harris, 2014). By using CCTV footage, traffic signalisation data, internet traffic data, flight and hotel reservation data, and pharmacy data, it creates so-called "narratives" or "alternative futures". While it was developed as a tool to forecast "potential events" in order to curb crime, or halt an epidemic, the government is now using it "to plan procurement cycles and budgets, make economic forecasts, inform immigration policy, study housing markets, and develop education plans for Singaporean schoolchildren" (Harris, 2014). The idea is then not just to: (1) focus on the behaviour of individuals suspected of crime or contagious diseases, but to "gauge the nation's mood"; (2) focus on crime, civil unrest, or health-related threats, but also to proactively design the social and economic policies of the country; (3) focus on threats, but on the strategic opportunities of the country. The idea of "smart cities" encapsulates the idea of total regulation, as their goal is not only to stimulate consumer spending to use natural resources more efficiently or to make traffic less heavy. The idea is centralise and empower the leading elite by sucking all data into one central AI tower or so-called "robotopticon" (Pasquale, 2017).

Big data and its discontents

While being seemingly more objective, knowledge and neutral language, big data, and algorithms carry several caveats.

The risk of de-identification

According to the European General Data Protection Regulation (GDPR), personal data is defined as "any information concerning an identified or identifiable natural person" (preamble, 26). In order to determine whether a natural person is "identifiable", account should be taken "of all the means reasonably likely to be used". Given the accelerating progress of big data analytics and data mining capabilities, less and less effort is needed to identify a person from scattered pieces of data. Such pieces do not tell much on their own, but they can be revealing and used to identify a person if taken at the aggregate level. They may create a complete mosaic of a person's life. The GDPR furthermore defines the "means reasonably likely to be used" by taking objective factors into consideration, such as the costs of and the amount of time required for identification, the available technology at the time of processing, and overall technological development. Such an open definition may be able to accommodate an elusive definition of personal data, but it makes clear that it is reasonable to expect that the existing personal data protection regime may absorb all sorts of data in the future. Such a definition of personal data can thus minimise the risk of de-identification due to increasing processing and data mining capabilities, which can turn seemingly impersonal data into personal data. However, the definition may progressively absorb other types of impersonal data and completely paralyse effective protection of personal data in the long run – when all data become personal data, compliance and enforcement become impossible or disproportionate and arbitrary.

Several cases concerning the risk of de-anonymisation have been documented in the press and literature. For instance, Sweeney analysed data from the 1990 US census and revealed that 87 per cent of the US population could be identified by just a ZIP code, date of birth, and gender. "[K]eeping data private, she insists, involves far more than just the removal of a name – and she's eager to prove that, with a quantitative, computational approach to privacy" (Perry, 2011). It is not only security flaws that can lead to a violation of personal data protection. What is new is the threat of de-anonymisation, which did not exist 30 years ago, when "no one was likely to pore through millions of records by hand to find patterns and anomalies". The threat thus demands the invention of new models for personal data protection to have effect given the increasing challenge posed by data de-anonymisation techniques. There are several solutions. For instance, Sweeny proposes mandatory insurance against major losses and compensation for data subjects according to the level of risk they incur, or what she calls a new "privacy-preserving marketplace" (Perry, 2011).

In summary, the personal data protection regime may become increasingly difficult to uphold due to a blurring of the lines between personal and impersonal data, which increases the risk of de-identification.

The needle-in-a-haystack problem and the ad infinitum hurdle

"I would argue that the most dangerous area in modern technology is not software that's too clever, but rather software that isn't clever enough" (Kroening, 2017). This argument basically claims that technology is not yet sufficiently perfected, but that self-learning machines may remedy the current deficiencies. "*The algorithm needs more time/data*", has been the argument of big data advocates, which means that inaccuracies resulting from algorithmic calculations are also deficiencies in their own right. In the domain of crime control, for instance, this entails that, in order to find a needle, e.g. a terrorist or [insert-anything], in a haystack, more data and more processing power are needed.

What we are witnessing here is the logic of ideology, i.e. in this case the ideology of big data. How the ideology works is discernible from the socialist experiment in the twentieth century. The elites in socialist countries always claimed that the socialist system had not yet been sufficiently perfected to overcome greed, wealth disparities, and other inequalities, which still de facto persisted in socialist countries. The existing political system – socialism – was not yet fully operational, yet, they boldly admitted, "we are on the right path to building a harmonious society". Socialism was presented as only an intermediate stage towards true liberation in the form of communism. However, the ideology should not be understood as something inherent in socialist regimes. After the fall of the Berlin Wall and the collapse of the socialist countries in Central and Eastern Europe, another ideology with same logic came to the fore. The new capitalist advocates claimed that "we are in the midst of transition; striving for a harmonious state of affairs should be given more time and commitment". They admitted that huge injustices were occurring during such transition, e.g. widespread corruption, looting by means of white-collar crime in broad daylight, and the appropriation of once publicly owned companies and natural resources, etc. But these were presented as deficiencies of the transition, when democratic institutions were still "immature": "Young democracies need support and encouragement", was the new guiding principle, and the support really came – in the form of a crash course in institution-building, privatisation programmes, and new lobbying businesses, as well as informal pressure directed against policy makers. Patience was again needed for the system to "mature". The fact that white-collar crime, corruption, land-grabbing, economic, and environmental crime, etc. are a part of every contemporary capitalist society was concealed. This normalised "systemic violence" (Žižek, 2008) was denied – the concealing gesture which remains at the core of ideology. What is at work in the transition countries is not a democratic deficit due to their "immature" democracy, but the natural state of affairs of capitalism.

The argument that software is not yet clever enough brings with it a similar concealing gesture: what is needed is more data, more storage, and more processing power. What is needed is patience and the motivation to make things "really" work: "smarter" algorithms have to be written, algorithms should monitor existing algorithms ad infinitum (Reynolds, 2017). The programmes will provide feedback in terms of rewards and punishments, and they will automatically progress and navigate the space of the given problem – known as "reinforcement learning" in machine learning. "I'm talking about AI that talks not to humans but directly to computer programs, creating software that codes, edits, and tests itself" (Kroening, 2017). The logics become caught up in escalating and circular decision-making loops ad infinitum.

In the crime and security domain, there is an insatiable desire to "throw more hay on the pile". For instance, when intelligence agencies strive to identify a "sleeping terrorist". But throwing more hay on the pile – more data to feed the "datasaur" – will not make that problem any easier (Schneier, 2006). It may also be that there are just too many needles in the haystack, and this makes it impossible to know which needles are problematic.

This logic persists in other algorithmic domains as well. When a Tesla autonomous car crashed into a truck when the car's cameras could not recognise a white truck against a bright sunlit background, the pro-autonomous car community amplified how the statistics were still in favour of autonomous vehicles. "Self-driving cars don't get sleepy, distracted, drunk, road-raged, and the many other things that cause about 90 per cent of crashes [...] robot cars could be a really important technology" (Lin, 2016). But as Lin shows (Lin, 2016), data scientists are concentrating too much on data and statistics. The lack of data may be the immediate reason for a crash and the new pieces of data related to this practical hurdle may prevent similar events in the future. But advocates of "more data is needed" neglect that while more data may make autonomous vehicles more safe, they still may not be acceptable enough for the public to embrace them for other reasons.

First, crash patterns will likely be different with robot cars – the people injured or killed will probably not be the same ones who would otherwise be victims (Lin, 2016; Lin, Abney, & Bekey, 2011). For instance, blind people that would be allowed to ride autonomous cars would be killed although they might not have been victims of a car crash today, because blind and other non-licensable drivers would not be operating a car today. A child riding in a car alone and involved in a crash might be something that the public at large will not accept. People may thus not accept algorithmic-driven technology because the car's cameras made a type of mistake that human drivers are very unlikely to make, e.g. the misperception of a white truck against a bright sunlit background for a white sky. Second, we are very uncomfortable with the possibility that a robot car might make a mistake that a human is less likely to make, while we are still willing to accept human mistakes, which are taken as a "road tax", but which robots will most likely never make. Third, we are very uncomfortable with the possibility that a robot car might choose to take a life, if it is the lesser of two unavoidable evils. This breaches the very concept of human dignity. Even in armed conflicts, where a combatant is liable to be legally killed at any time, there is "a right to life", which means a right not to be killed arbitrarily, unaccountably, or otherwise inhumanely (Lin, 2015; Lin et al., 2011) – inhumanely may also mean with an autonomous weapon without a human in the decision loop.

The eternal past: self-fulfilling prophecies and the vicious circle effect

Computers can only be tasked with making "inductive" predictions based on past experiences. The future they predict is a continuation of the past data and not behaviour itself. The future is then calculated on the basis of already selected facts about facts. Predictions can be more accurate in cases where "reality" does not change dramatically and where the data collected reflects "reality" as closely as possible. First, by default, criminality is never fully reported. The dark figure of crime is a "black box" that can never be properly encompassed by the algorithms. Second, crime is a normative phenomenon, i.e. it depends on human values, which change over time and place. Algorithmic calculations can thus never be accurately calibrated given the initial and changing set of facts or "reality".

In the crime and security domain, this means that predictive policing algorithms rely on records generated by police forces themselves – and not crime as it actually is. The machines then calculate the probability of future police records and can only reinforce the behaviours we have always known. They may thus amplify the prejudicial behaviour of police that led to the data that is being acted upon in the first place (Smith IV, 2015). In police work, self-fulfilling prophecies occur in the form of "overpolicing", i.e. when the police increase their presence in the same places they are already policing, and thus in turn uncover more crime, which fosters the need to police the areas or the targeted communities even more. Such cases have been widely acknowledged in inspections of specific police algorithms, such as in a study on the Oakland predictive policing programme. The study revealed how PredPol amplifies racially biased policing in racially segregated communities (Smith IV, 2016). PredPol would have sent Oakland officers to mostly black neighbourhoods despite actual drug crime being evenly distributed across racial lines. This shows the power or "performativity" of data (Raley, 2013). Just because the data subject is thought to have an inclination to certain behaviours that are not yet evident, data leads to some action directed against a subject. It is the power of data that produces suspects, i.e. the "effect of producing that life, that body as a [...] suspect" (Raley, 2013).

Blurring probability, causality, and certainty

Risk assessments yield probabilities, not certainties. They measure correlations and not causation and do not reflect objective truth, as "data fundamentalists" – a term coined by Crawford (2013) – would claim. They are similar to regression analysis, which offers extremely partial and narrowly focused predictions. While they may be very good at finding correlations, they cannot judge whether or not the correlations are real or ridiculous (Rosenberg, 2016). Correlations thus have to be examined by a human, they have to be ascribed a meaning. Similarity to past cases does not automatically yield meaning to future cases; for instance, an offender's acts cannot be explained merely by past cases – the numbers of the past do not automatically speak for the present. In other words, ascribing weight to past correlations in the form of oscillating between the past and present is an inherently political/value-based process. For instance, at what probability of the risk of reoffending should a prisoner be granted parole? Whether this ought to be 40 per cent or 80 per cent is an inherently political decision based on the social, cultural, and economic conditions of a given society, e.g. full prisons may lead to a mild parole policy, while governance through fear and a burgeoning prison industry may lead to a harsher parole policy. The intertwining of policies in a given society is crucial for attributing numbers a specific policy meaning, and Wacquant showed how the neoliberal economic agenda has triggered harsher sentencing and decreased the frequency of parole being granted – Smith's "invisible hand" of the free market is already wearing an iron glove (Wacquant, 2009). A person with a higher risk of reoffending may have been granted parole 20 years ago, while the policy changes of the neoliberal era dictate the opposite decision. Data analytics translates differently in different times due to the changed political, social, and cultural landscape. There is no Archimedean point (*Punctum Archimedis*) from which to ascertain the "correct" translation of numbers into decisions, i.e. there is no hypothetical ideal vantage point from which accurate mathematical predictions of future human behaviour can be seen: a 60 per cent chance of reoffending has no necessary implications as regards sentencing policy.

Statistical modelling: from dirty data to runway algorithms

The dispute between Wittgenstein and Turing about the nature of numbers, as so lucidly presented by Amoore (2014), shows how mathematics can be used for security calculations. For Wittgenstein, "mathematics 'makes no predictions', but instead is a form of grammar: 'taken by itself, we shouldn't know what to do with it; it's useless. But there is all kind of use for it as part of a calculus'" (Amoore, 2014). However, as Amoore states, Turing insisted on the capacity of numbers: "one can make predictions". Against the backdrop of such a contradictory understanding of mathematics as ranging between intuition and ingenuity, Amoore claims how contemporary "rule-based" security decisions involve both intuition and ingenuity:

> the proliferation of what have come to be known as "risk-based" and "rules-based" security decisions is just such an intuitive bridging of the gaps in available data and an ingenuity of algorithmic rules to make this routine replicable into the future. A risk-based security technique is based on a set of decision procedures through which a final calculation is produced – 'Is this factor present?', 'Is this variable co-present?', and so on. Though it is a set of association rules – a "rules-based" programme – that allows for the risk calculation to be made with ingenuity, it is intuition that supplies the identification of patterns.
>
> (Amoore, 2014: 427–428)

The data do not speak for themselves. In the language of Turing, intuition has yet to provide meaning to the calculus.

Almost every factor that forms the basis for assessment also has its own history. This includes sociality by default: "Data is not a natural resource but a cultural one" (Gitelman, 2013). For instance, an individual's history of arrests carries meaning, e.g. it is problematic in racially divided communities, where low-income and non-white residents are arrested more than others. If an individual's history of arrests is a factor in risk assessment, this distorts the calculus as it cleanses the data of social context – i.e. that some people are arrested and overpoliced more than others. Similarly, using convictions indiscriminately cannot be "objective". The ACLU has demonstrated, for instance, how blacks are more likely than whites to be convicted of marijuana possession, even though they use the drug at rates equivalent to whites in the USA (ACLU Foundation, 2013). Every factor that forms the basis for the assessment has its own history and includes sociality by default.

There are at least two challenges in algorithmic design and application: (1) compiling statistical models and building algorithms always requires decisions that are made by humans and (2) algorithms may also take an unpredictable path to reach their objectives.

The *first* question relates then to data – not only the nature of mathematics and data per se, as presented above, but more specifically which data is taken in and used and which data is left out of the calculus, i.e. how data is cleaned and prepared. Furthermore, different sorts of data are not of the same quality, e.g. social media posts are not as trustworthy as official records, such as property records. In such a case, it does not help if more data is entered into the data-crunching machine; if the data is of poor quality, then "garbage in garbage out".

The process of preparing data is inherently political in the sense that the methodological decisions are made by humans. Data needs to be generated, protected, and interpreted (Gitelman, 2013). For instance, sentiment analysis deduced from Twitter use will only take into account a specific population that comes from specific social strata; e.g. urban middle class users, who are not a representative cross section of society. For instance, when Slovenian police were analysing Twitter accounts in the aftermath of the "Occupy Movement" in Ljubljana in October 2011, it could only reach part of the population as the number of Twitter accounts in Slovenia amounts to 10 per cent of the whole population, and just one sixth of the accounts are used daily (Valicon, 2016). The so-called "signal problem" (Crawford, 2013), where little or no data comes from a certain segment of society must also be taken into account. The very use of a given technology is, additionally, always in flux: "any divide in accessing digital technology is not a one-time event but a constantly moving target as new devices, software, and cultural practices emerge" (Crawford, 2013).

The process of preparing data encompasses decisions as to which data is selected and how it is cleaned and prepared. This process can contain mistakes, biases, and attitudes, or simply reflect how problems are framed or even the feelings of the data scientists. Crime reports and other statistics gathered by the police are not, as mentioned above, an accurate record of all the crime that occurs in a community; instead, they are partly a record of law enforcement's responses to what happens in a community. Predictive policing thus risks fuelling a cycle of distorted enforcement (Robinson & Koepke, 2016). The "information revealed by big data analysis does not offer an impartial overview of any subject matter and is only as reliable as the underlying data permits." (European Parliament, 2017).

The interpretation of results is not straightforward as well, as it is inherently affected by human knowledge of the analysed domain and data. What in fact is calculated is difficult to interpret by data scientists without specific knowledge of the origin of the data. Typically, data scientists have to work alongside a domain-specific scientist in order to ascribe meaning to the calculated results. "We give numbers their voice, draw inferences from them, and define their meaning through our interpretations" (Crawford, 2013). The meaning of the calculated results will often vary over time and place. Just as biases can slip into the design of a statistical model, they can also slip into the interpretation – results are interpreted according to political orientations, values, and framings imposed by the interpreter.

The *second* question relates to the question of how algorithms can take an unpredictable path to reach their objectives as they are not neutral but rather reflect choices about data, connections, inferences, interpretations, and thresholds for inclusion that advance a specific purpose (Dwork & Mullingan, 2013). For instance, if an employer does not want to hire candidates planning to have children due to the burden of future maternity leave, but knows that Catholic women tend to have more children, religious belief could be calculated from other innocuous data. While the religious belief of candidates may not be directly included in the calculation in the screening process, a programme may attribute a lower score to candidates by calculating religious belief from purchase data, e.g. purchasing relatively more eggs and ham before Easter. While this is an example of an intentionally discriminatory design of an algorithm, artificial intelligence experts designing autonomous (self-driving) vehicles have warned how "smart vehicles" could unintentionally change the game. For instance, they might not only optimise their route in order to adapt to the traffic light regime, but might even attempt to optimise matters by interfering with the traffic infrastructure and traffic signalling devices. Similarly, in a Pulitzer Prize-winning investigative work, ProPublica showed biases in probation algorithms (Angwin, Larson, Mattu, & Kirchner, 2016): blacks are almost twice as likely as whites to be labelled a higher risk but to not actually reoffend, while the opposite mistake occurs among whites. They are much more likely than blacks to be labelled lower risk but go on to commit other crimes. This is not a direct calculation based on race, but is calculated by means of proxies that disproportionally hit racial groups (Harcourt, 2015).

Risks beyond civil liberties

Many personal data protection scholars and governing bodies (cf. Article 29 Working Party [2014]; the International Working Group on Data Protection in Telecommunications [2014]) have pointed out how big data analytics may severely infringe civil liberties (e.g. privacy, personal data protection, and the prohibition of discrimination). For instance, in its 2017 Resolution of 14 March 2017 on the fundamental rights implications of big data (2016/2225 [INI]), the European Parliament acknowledged that big data can lead to social sorting and indirect discrimination against "groups of people with similar characteristics, particularly with regard to fairness and equality of opportunities for access to education and employment, when recruiting or assessing individuals or when determining the new consumer habits of social media users" (European Parliament, 2017).

However, in addition to the impact of big data analytics on fundamental liberties, predictive technologies' impacts may have distorting effects on the fundamental cornerstones and architecture of liberal democracies, e.g. regarding the principle of the division of power and the limitation of political power by the rule of law. How big data can interfere with democratic processes can be discerned from the judgment of the European Court of Human Rights (EctHR) in *Roman Zakharov* v. *Russia* (47143/06, from 4 December 2015) and from the 2016 presidential elections in the USA.

The ECtHR case revealed that the law enforcement and intelligence agencies in Russia had direct access to all mobile phone data in Russia. This clearly endangers fundamental liberties, as the ECtHR decided in the case. However, the power to intercept, store, and mine such an amount of data on every individual – the penetration of mobile phones is extraordinary in surveillance capitalist societies, and the number of mobile phones far exceeds the number of inhabitants (in Russia there are 155 phone numbers per 100 inhabitants) (Wikipedia, 2017) – can lead to a distortion in democratic processes, elections, and the system of check and balances. With such data at its disposal, the government could potentially identify and track a particular individual deemed interesting by the government elite. It could conduct more profound in-depth analysis of the public's mood ("sentiment analysis") and identify "hotspots", i.e. opposition leaders and groups disseminating dissent.

The accumulation of vast amount of location and traffic data in the form of indiscriminate bulk surveillance is in fact practised by all major countries and not only Russia. For instance, the "Five Eyes" countries operate similar programmes regarding internet traffic, as so distinctively revealed by Edward Snowden's classified information leak in 2013. The coupling of proprietary data and governance in society has strikingly surfaced in the public consciousness in the 2016 presidential elections in the USA. On the one hand, Republicans were using the big data company Cambridge Analytica to target voters, while the Democrats were using an algorithm called ADA, which was supposed to help the Democratic camp win, but it eventually failed. One may only speculate, but perhaps the ADA algorithm would have been branded as a "president-making algorithm", which could be then rented to presidential candidates around the world. The algorithm was leading "every strategic decision Clinton's aides made, including where and when to deploy the candidate and her battalion of surrogates and where to air television ads – as well as when it was safe to stay dark" (Wagner, 2016). The Chinese government operates similar programmes, as comparative studies on the USA and China have shown (Gagnon, 2008).

What is new in digitised surveillance capitalism is that insights into digital communication networks, e.g. social media and public telecommunication networks, can be measured and also disrupted. The spread of alternative ideas and political opposition can be eliminated before it grows into a significant political alternative. The Facebook experiment with massive-scale emotional contagion of its users (Kramer, Guillory, & Hancock, 2014) clearly demonstrated how today there exist powerful tools for inducing or disrupting the spread of ideas. Such tools used for "emotional contagion" may in the future be used to produce "political contagion" amongst the public at large. It is not clear whether countries are using such data to scan the population for such political ends, but they are clearly using social media sentiment analytical tools to curb social unrest and public disorder. Examples come from many countries in the form of policing the web and scanning social media accounts. For instance, even internationally less significant players, such as Slovenia, used – as the police publically admitted – Twitter analysis to scan the participants and leaders of the "Occupy Movement" in Ljubljana in October 2011. The details of the programme have not been revealed in order to protect the so-called "tactics and methods of police work", and the police claim that this was used solely for "operative and not profiling purposes".

Why do we need this book?

Providing a "working definition" of the elusive concepts of big data and related algorithmic tools may be a Sisyphean task. The book instead strives to look at the social, economic, and political implications, goals, and aspirations of big data analytics in social and crime control. The mathematical predictions and reasoning used in an increasing number of social domains make the study of big data and algorithms inspiring and frightening at the same time. It is not only the idea of the exponentially increasing computer capabilities heading towards the point of "technological singularity" – the hypothesis that the invention of artificial superintelligence will abruptly trigger runaway technological growth (Wikipedia, 2017b), as the Silicon Valley elite eagerly warn – that makes this book a necessity. The idea is that there are socially destructive consequences of big data and automated decision-making systems already at work and unfolding in surveillance-based capitalist societies in the form of discrimination against the less affluent and less powerful parts of the population.

The contributors to this book feel that "big data" has been granted too much agency and too much power too quickly. Big data is shifting power relations in several domains, from banking (O'Hara & Mason, 2012), insurance (Meek, 2015) (Ambasna-Jones, 2015), education (Ekowo & Palmer, 2016; Selingo, 2017), and employment (Cohen et al., 2015), to the domains of control and security (McCulloch & Wilson, 2015). It is also increasingly interfering with democratic political processes (Confessore & Hakim, 2017). Its predictive potentials have become an attractive method of predicting human behaviour in too many contexts: It has been used to combat financial risks, fortify healthcare, conquer spam, improve the fight against crime, and boost sales (Siegel, 2013). It has been vested with a great deal of power, while at the same time presented as an objective, value-free scientific tool that requires no transparency or auditing – indeed, no further explanation whatsoever. It has become a projection screen for our desire to predict and colonise the future – to eliminate all the risks to our well-being, but only for those who can afford a data scientist – "the sexist job of the twenty-first century" (Davenport & Patil, 2012). Big data may be bringing about a revolution that "will transform how we live, work, and think" (Mayer-Schönberger & Cukier, 2013), but this revolution will not occur because of big data itself, but because of the specific social, cultural, political, and economic imperatives in our society that allow such technology to flourish to the detriment of other types of knowledge and social practices.

About the book

The book is divided into five parts that touch upon different aspects of crime and social control in the light of big data and predictive algorithmic analytics. It focuses on changes in knowledge production and the manifold sites of contemporary surveillance, ranging from self-surveillance to corporate and state surveillance. It tackles the implications of big data and predictive algorithmic analytics for social justice, social equality, and social power, i.e. concepts at the very core of crime and social control. The leading questions include: Who benefits from big data and who loses? What type of knowledge and narrative is being promoted by big data and algorithmic language, and what type of knowledge and narrative is being supressed? Who is in the focus of the "algorithmic gaze" and who escapes its gaze? How are traditional actors in criminal justice systems responding to the new "algorithmic service"? These and other questions that big data and algorithmic analytics trigger in the domain of social and crime control are addressed from legal, criminological, sociological, and psychological perspectives. There is a mix of quantitative and qualitative approaches to illuminate the different research problems and to nuance the analysis of the topic. *Qualitative* approaches to and a *critical* perspective on big data and predictive analytics form the core of the book.

Part I (Introduction) introduces the subject and presents the definitional challenges of big data and predictive algorithmic analytics, followed by the origins of big data's "paradigm shift", and the caveats inherent in big data and the algorithmic calculus in the domain of crime and social control. Part II (Automated social control) paints the landscape of big data and social control from broader and critical theoretical perspectives. First, it draws on critical theory of surveillance and tackles the disparities big data triggers – e.g. how automated surveillance is too often directed at the "usual suspects", such as conventional criminals and minority communities, while "high net worth individuals" are benefiting from privacy and encrypted capital flows. Second, by applying psychological and psychoanalytical insights, this part tackles the ignorance related to the new self-surveillance practices and "forced choices" data subjects are offered for their digital life. Third, by drawing on critical criminological theory, this part tackles the processes of the "commodification" and "datafication" of data subjects in postmodern capitalism in the light of "class struggle". Part III (Automated policing) focuses on the origins, implicit assumptions, economy, and consequences of predictive policing, while Part IV (Automated justice) delves into the supposedly more "objective" and "unbiased" decision-making tools offered to criminal justice system actors and provides insight into the established antecedents of the "datafication" of criminal justice processes, ranging from the "moral statistics" of the ninteenth century to "new penology", "actuarial justice", and intelligence-led policing. Finally, Part V (Big data automation regulation) focuses on the legal boundaries of data processing from the personal data protection and criminal procedure points of view, with the last part examining possible avenues to curb economic cyber espionage – state-sponsored economic big data theft – from the international law perspective.

Readers may also approach the book through other trajectories: readers with an interest in specific *forms of crime* should examine Chapter 2 (author: Pasqaule), for insights related to financial crime and tax havens, and Chapter 10 (Veber and Kovič Dine) for discussion on cyber crime and cyber espionage. Those interested in changes in *subjectivity and narrative* brought by big data and algorithmic analytics in the criminal justice system should turn to Chapter 5 (Andrejevic), Chapter 7 (Završnik), and Chapter 8 (Plesničar and Šugman Stubbs) – these chapters approach the de-subjectivation effect of big data prosthetics from different but complementary theoretical angles. Readers interested in *critical approaches to surveillance and privacy* may choose to focus their attention on Chapter 2 (Pasqaule), Chapter 3 (Salecl), Chapter 4 (Kanduč), and Chapter 5 (Andrejevic). The topic of *policing* is tackled in great depth in Chapter 5 (Andrejevic), Chapter 6 (Wilson), and Chapter 7 (Završnik), while there is a focus on *criminal justice* in Chapter 7 (Završnik), Chapter 8 (Plesničar and Šugman Stubbs), and Chapter 9 (Gorkič). The common thread of the *legal* implications and regulation of big data runs through Chapter 2 (Pasquale), Chapter 9 (Gorkič), and Chapter 10 (Veber and Kovič Dine).

In Part I (Introduction – Chapter 1: "Big data: what is it and why does it matter for crime and social control?"), Aleš Završnik presents the definitional challenges and origins of big data, and what they entail for crime and social control. It focuses on the semantic novelties of big data and algorithmic analytics, which are blurring contemporary regulatory boundaries, undercutting the safeguards built into regulatory regimes, and abolishing subjectivity and case-specific narratives. It continues by presenting the knowledge-production change provoked by big data and the subsequent changes this triggers in terms of social power (im)balances. The author claims that we are witnessing changes in the governance model in several domains, wherein big data and algorithms feature similar inherent deficiencies, which are identified and analysed in great detail.

Part II (Automated social control) begins with Chapter 2: "Paradoxes of privacy in an era of asymmetrical social control", in which Frank Pasquale adds a third dimension to the "privacy v. security" debate by bringing in the relative wealth and privilege of targets of surveillance. Pasquale claims that the wealthy are often the "prime beneficiaries of obfuscatory law and technology, while the lives of others are all too often an open book." By drawing on critical theory of surveillance and an analysis of international tax evasion, Pasquale sketches an egalitarian approach to data policy and governance. Pasquale paints a very worrying picture of not only the so-called "algorithmic divide", where the affluent benefit from personalised services and algorithms, while the poor may only receive an automated response, but also of how new technologies are entering a specific cultural and societal space characterised by social inequality and the unequal division of wealth. Big data analytics, encryption, etc. are too often used as means of deepening and widening social exclusion, influencing social inequality, and weakening social cohesiveness. By presenting the dynamics of financial crime, especially certain tax evasion schemes, Pasquale clearly shows the paradox of privacy and the hypocritical approach of many governments when navigating between privacy and transparency. While the "homeland security apparatus has successfully lobbied for extraordinary funding for tanks for rural towns, monitoring Occupy activists, and a surfeit of flat screen televisions for fusion centres," claims Pasquale, the heart of terror financing stays protected by privacy regimes. "The wealth defence industry, devoted to exploiting secrecy jurisdictions and tax havens for its dubious clients, remains as strong as ever, despite growing knowledge of just how destructive its methods can be."

In Chapter 3: "Big data – big ignorance", Renata Salecl tackles the current ideology of being "successful", "happy", etc. behind the wearable computing and self-surveillance movement. By drawing on Lacanian theoretical psychoanalysis and recent psychological studies into negation, Salecl draws crucial distinctions between ignorance, repression mechanisms, and misrecognition. Salecl particularly focuses on the legalistic concept of "informed consent" in the domain of personal data protection, and questions the meaning of the lengthy pages of information provided to data subjects which are supposedly intended to result in their "informed" decisions and/or consent, although users most typically do not read, understand, or apply this information and thus "choose" to stay ignorant. The "informed consent" regime merely allows digital behemoths to collect, process, and sell their data to an unprecedented degree, to the detriment of the data subjects. Why, despite all the media coverage about the "spread of surveillance", do people not seem to care about being tracked, "datafied", "commodified", and sold for a profit? Salecl claims that users are in fact only in a position to provide "forced consent". In surveillance capitalism, the often ridiculed robber's choice "Your money or your life!" translates into "Your data or your digital death!". Furthermore, Salecl shows how the legal regulation of informed consent enables digital corporations to transfer the responsibility to users. Users are to be blamed for damage arising from data misuse/abuse and invasions into their private lives; they are solely liable for the limited life chances and life choices resulting from algorithmic social sorting effects. In this manner, Salecl shows how the structural tensions of digital capitalism and its "systemic" violence become individualised.

In Chapter 4: "Machines, humans, and the question of control", Zoran Kanduč situates the new information technology in the class struggle perspective. Kanduč claims that IT technology is not actually a liberating tool for the working class, but, on the contrary, a new tool for subordinating the masses, "a means for 'all too old' purposes, i.e. the exploitation and oppression of structurally subordinated workers." From a broader perspective, not only pertaining to big data and predicative analytics, but technology ranging from cars, consumer commodities, personal computers, etc., Kanduč theorises possible resistance against scientific and technological progress, because the current notion of progress is presented as being apolitical and without sufficient social, legal, and moral control. In a similar vain as Pasqaule, Kanduč claims that privacy advocates too often neglect the paradox of privacy – that the controlling power of repressive state apparatuses should not be exaggerated and the criticism should instead be directed at the dictatorship of the financial markets, multi- and transnational corporations ("global players"), and the institutions of the "supra-national state of capital", which is too often protected by the walls of privacy and secrecy.

Part III (Automated policing) focuses on automated, or algorithmic, policing and contains chapters that take as their point of departure two different theoretical backgrounds. In Chapter 5: "Data collection without limits: automated policing and the politics of framelessness", Mark Andrejevic discusses how big data has transformed policing with the promise of not only making policing more efficient, but that information might contribute to the goal of pre-emption. But in an era of an informational glut, this pre-emption may be achieved not only by total information capture, but also by means of technologies for making sense of it and putting it to use. Policing is now more about information than weaponry or the "weaponisation of data". While theorising this state of post-panopticism, Andrejevic introduces the concept of framelessness, i.e. the increase in the total information awareness without borders seen in predictive policing styles with a strong tendency to push beyond historical crime data. The concept applies to disappearing boundaries, e.g. of what type or amount of data should be collected and processed, and what types of data should not be collected, or between individual behaviours and spatial patterns of crime. These boundaries have fallen. The collection of "frameless" data does not rule out anything in advance. The chapter offers several important insights into how new policing technologies and other technologies automating sight subtract the human being from the system. This "subtraction" is part of the policing style wherein machines find targets or where machines, such as wearable cameras, when enhanced with face recognition and other similar software, can react and preemptively act before a human police officer would. The chapter shows how data-driven policing is only one part of this framelessness. The other pertains to the targets since anyone can pose a potential risk. Andrejevic discusses the resurgence of the category of the "undisciplinable subject" that emerged in the post-9/11 era and the subsequent rereading of criminality through the lens of terrorism: criminals are "other" and "alien", incapable of internalising or abiding by the norms and values of liberal democracy. While tackling the figure of the terrorist, Andrejevic thus suggests:

> subjection to panoptic surveillance is at once too much and not enough – too much because they do not need to be told they are being monitored (this would not have the desired disciplinary effect) and not enough because they must be monitored all the time, as comprehensively as possible. [...] Post-panopticism invokes the goal of framelessness: that of total information capture.

In Chapter 6: "Algorithmic patrol: the futures of predictive policing", Dean Wilson presents the historical antecedents and cultural embeddedness of predictive policing. The author shows the cultural refocusing from causes of crime to the surface of crime and the obsession of society with correlation instead of causality – there has been a reformulation of the central questions in criminology from *why* to only *what*, i.e. not why a crime was committed and how to address the root causes, but merely what crime was and will be committed. The author furthermore shows how the CompStat paradigm stems from the "broken windows theory", which means that it focuses only on minor crimes, and how innovations, such as PredPol, aspire to go beyond minor criminality, but without clearly presented proof of crime reduction success. In an eloquent narrative, Wilson touches upon all major contemporary policing changes, in particular the technological enhancement of policing, which has an aura of "progress", the more recent imbalance between the fighting crime v. social service functions of policing, and the militarisation of the police. Wilson shows how tools that may seem like objective means for responding to crime are never neutral – they perpetuate existing power relations, which too often translate into the formula: "the rich get richer and the poor get prison". What may be surprising is that algorithmic prosthetics not only discriminately target individuals and communities, but also significantly impact police work and its alienation from the community.

Wilson also points to increasing police interest in social-media big data in the form of social-media intelligence. By situating policing in the technophilia context, Wilson shows how predictive policing may erode trust in the police externally, i.e. vis-à-vis policed communities, and internally in the form of decreased self-perception of their legitimacy.

Part IV (Automated justice) bridges theorising big data and predictive analytics in crime control by steering the debate from policing to courts and other actors in criminal justice systems, e.g. parole boards. In Chapter 7, Aleš Završnik tackles the well-established antecedents of the datafication of criminal justice processes, such as "moral statistics", actuarial justice, and crime mapping. The chapter presents how these antecedents may be similar to big data on the surface, but are in fact significantly different. It shows how big data attaches to penal power, in a similar vein as statistics and psychiatry attached to the science of crime in the nineteenth century. In the central part, several programmes that are already being used or tested in either policing or criminal justice systems are presented and critically assessed. The author situates "automated policing" and "automated justice" in the neoliberal "datafication of social control" context. In Chapter 8, Mojca M. Plesničar and Katja Šugman Stubbs focus on the psychological factors of decision-making in the courtroom and draw a crucial distinction between what they call "harmful subjectivity" and "constructive subjectivity". They show how the positivists' efforts to minimise subjectivity in the criminal justice decision-making process, e.g. to completely minimise human bias, mistakes, etc. in order to achieve impartial and objective outcomes, may in fact be harmful. In other words, while the positivists pursue legitimate interests, the means of reaching such a perfect state are flawed. Instead, Plesničar and Šugman Stubbs suggest, subjectivity should not be completely abolished, and substituted for by computers, as advocated by big data entrepreneurs, but constructive elements of subjectivity should be embraced. The authors advocate the view that similarly to the sentencing tables that criminal justice professionals used in the pre-digital era, big data analytics and computerised decisions in judicial settings "reflect a deeply ingrained distrust of the professional knowledge commonly attributed to judges". The authors show how the algorithms used in a courtroom are not able to live up to their promises, at least not yet. They optimistically conclude, inter alia, that also algorithms may in the future be designed in such a way so as to emulate good and constructive subjectivity.

Part V (Big data automation limitations) delves into the legal conundrums of big data, specifically in the criminal law and international public law context. In Chapter 9: "Judicial oversight of the (mass) collection and processing of personal data", Primož Gorkič focuses on an analysis of the need for judicial oversight of the collection and processing of traffic communication data and DNA-based data. The central question guiding the analysis is whether and why legislators should provide some form of judicial oversight of government personal data processing in these two cases – i.e. DNA and traffic communication data. Gorkič approaches the question by comparing the European approach, as delineated in the jurisprudence of the ECtHR and the European Court of Justice, and the jurisprudence of the USA. The author posits the view that the divide in privacy conceptualisations dictates the methodologies courts employ in establishing whether the mass collection and processing of personal data should be subject to judicial oversight. While the European "privacy-as-personhood approach", i.e. the dignity-centred approach, reveals the added value of judicial oversight, the "conduct-oriented approach" of the US courts finds no need for *ex ante* or *ex post* judicial oversight. Gorkič concludes that in cases involving increasingly clandestine surveillance practices, judicial oversight is the next best alternative to an individual's participation in the decision-making process concerning how her or his personal data should be collected and processed.

The concluding Chapter 10: "Big data and economic cyber espionage: an international law perspective", focuses on state-sponsored theft of economic big data and the possible avenues for regulating cyber espionage, which seems to be destabilising the global economy by enabling "the greatest transfer of wealth in history." Maruša T. Veber and Maša Kovič Dine first define economic cyber espionage and distinguish it from industrial espionage – while the former is sponsored by states, the latter pertains to private sector entities. After presenting the methodological hurdles in establishing the scope of economic cyber espionage, the authors focus on a legal analysis of the three alternative international legal avenues for countering state-sponsored theft: (1) bilateral agreements addressing the question of the legality of economic cyber espionage; (2) general international law rules, such as the principle of non-intervention; and (3) trade policy tools. The latter is a provocative and creative way to address big thefts by way of relying either on the World Trade Organisation Agreement, the Trade-Related Aspects of Intellectual Property Rights (TRIPS) Agreement, or on free trade agreements with provisions related to trade secrets. The authors posit the view that economic cyber espionage differs substantially from other types of espionage, which may subsequently lead to stronger prohibitions regarding state-sponsored economic cyber espionage at the international level.

References

ACLU Foundation. (2013). *The war on marijuana in black and white*. New York. Retrieved from www.aclu.org/report/report-war-marijuana-black-and-white?redirect=criminal-law-reform/war-marijuana-black-and-white.

Ambasna-Jones, M. (2015, 3 August). The smart home and a data underclass. *Guardian*. Retrieved from www.theguardian.com/media-network/2015/aug/03/smart-home-data-underclass-internet-of-things.

Amoore, L. (2014). Security and the incalculable. *Security Dialogue*, *45*(5), 423–439.

Anderson, B., & Horvath, B. (2017, 2 September). The rise of the weaponized AI propaganda machine. *Scout*. Retrieved from https://scout.ai/story/the-rise-of-the-weaponized-ai-propaganda-machine.

Anderson, C. (2008, 23 June). The end of theory: The data deluge makes the scientific method obsolete. *Wired*. Retrieved from www.wired.com/2008/06/pb-theory/.

Andrejevic, M. (2013). *Infoglut: How too much information is changing the way we think and know*. Abingdon, Oxon; New York, NY: Routledge.

Angwin, J., Larson, J., Mattu, S., & Kirchner, L. (2016, 23 May). Machine bias: There's software used across the country to predict future criminals. And it's biased against blacks. *ProPublica*. Retrieved from www.propublica.org/article/machine-bias-risk-assessments-in-criminal-sentencing.

Blackie, R. (2017, 2 July). Did Cambridge Analytica win the election for Trump? Retrieved from http://robblackie.com/did-cambridge-analytica-win-the-election-for-trump/.

Bradfield, A. L., Wells, G. L., & Olson, E. A. (2002). The damaging effect of confirming feedback on the relation between eyewitness certainty and identification accuracy. *Journal of Applied Psychology*, *87*(1), 112–120.

Caliskan-Islam, A., Bryson, J. J., & Narayanan, A. (2016). Semantics derived automatically from language corpora necessarily contain human biases. Retrieved from http://randomwalker.info/publications/language-bias.pdf.

Captain, S. (2015, 28 September). Hitachi says it can predict crimes before they happen. *Fast Company*. Retrieved from www.fastcompany.com/3051578/elasticity/hitachi-says-it-can-predict-crimes-before-they-happen.

Cohen, J. E., Hoofnagle, C. J., McGeveran, W., Ohm, P., Reidenberg, J. R., Richards, N. M., et al. (2015). *Information privacy law scholars' brief in Spokeo, Inc.* v. *Robins* (SSRN Scholarly Paper No. ID 2656482). Rochester, NY: Social Science Research Network. Retrieved from http://papers.ssrn.com/abstract=2656482.

Confessore, N., & Hakim, D. (2017, 6 March). Data firm says "secret sauce" aided Trump; many scoff. *New York Times*. Retrieved from www.nytimes.com/2017/03/06/us/politics/cambridge-analytica.html.

Crawford, K. (2013, 1 April). The hidden biases in big data. *Harvard Business Review*. Retrieved from https://hbr.org/2013/04/the-hidden-biases-in-big-data.

Davenport, T. H., & Patil, D. J. (2012, 1 October). Data scientist: The sexiest job of the 21st century. *Harvard Business Review*. Retrieved from https://hbr.org/2012/10/data-scientist-the-sexiest-job-of-the-21st-century/.

de Kort, Y. (2014, May). Spotlight on aggression. *Intelligent Lighting Institute, Technische Universiteit Eindhoven*, *1*, 10–11.

Desrosières, A. (2002). *The politics of large numbers: A history of statistical reasoning*. Cambridge, MA: Harvard University Press.

Dowd, M. (2017, 11 January). Peter Thiel, Trump's tech pal, explains himself. *New York Times*. Retrieved from www.nytimes.com/2017/01/11/fashion/peter-thiel-donald-trump-silicon-valley-technology-gawker.html.

Dwork, C., & Mullingan, D. K. (2013). It's not privacy, and it's not fair. *Stanford Law Review*, *66*(35). Retrieved from www.stanfordlawreview.org/online/privacy-and-big-data-its-not-privacy-and-its-not-fair/.

Ekowo, M., & Palmer, I. (2016). *The promise and peril of predictive analytics in higher education*. New America. Retrieved from www.newamerica.org/education-policy/policy-papers/promise-and-peril-predictive-analytics-higher-education/.

Franko Aas, K. (2005). *Sentencing in the age of information: From Faust to Macintosh* (1st edn). London: Routledge-Cavendish.

Franko Aas, K. (2006). "The body does not lie": Identity, risk and trust in technoculture. *Crime, Media, Culture*, *2*(2), 143–158.

Gagnon, B. (2008). Cyberwars and cybercrimes. In *Technocrime: Technology, crime and social control* (pp. 46–65). Cullompton: Willan.

Galič, M. (2017). Živeči laboratoriji in veliko podatkovje v praksi: *Stratumseind 2.0 –* diskusija živečega laboratorija na Nizozemskem. In: A. Završnik (Ed.), *Pravo v dobi velikega podatkovja*. Ljubljana: Institute of Criminology at the Faculty of Law.

Gartner, I. (2012, 25 May). What is big data? – Gartner IT glossary – big data. Retrieved 8 March 2017, from www.gartner.com/it-glossary/big-data/.

Gitelman, L. (Ed.). (2013). *"Raw data" is an oxymoron*. Cambridge, MA.: MIT Press.

Harcourt, B. E. (2015). Risk as a proxy for race. *Federal Sentencing Reporter*, *27*(4), 237–243.

Harris, S. (2014, 29 July). The social laboratory. *Foreign Policy*. Retrieved from http://foreignpolicy.com/2014/07/29/the-social-laboratory/.

Hey, T., Tansley, S., & Tolle, K. M. (2009). *Jim Gray on eScience: A transformed scientific method.* Retrieved from http://scholar.google.com/scholar?cluster=6187259227944398156&hl=en&oi=scholarr.

IBM. (2016). The four V's of big data. Retrieved 20 January 2017, from www.ibmbigdatahub.com/infographic/four-vs-big-data.

Information and Privacy Commissioner of Ontario (2017). Government and big data: Privacy risks and solutions, 26 January 2017, lecture. Retrieved from: http://livemedia.biz/ipc2017.html.

Innocence Project. (2016). Eyewitness misidentification. Retrieved 29 August 2016, from www.innocenceproject.org/causes/eyewitness-misidentification/.

Kassin, S. M. (2008). Confession evidence commonsense myths and misconceptions. *Criminal Justice and Behavior*, *35*(10), 1309–1322.

Kassin, S. M., & Kiechel, K. L. (1996). The social psychology of false confessions: Compliance, internalization, and confabulation. *Psychological Science*, *7*(3), 125–128.

Kerr, I., & Earle, J. (2013). Prediction, preemption, presumption: How big data threatens big picture privacy. *Stanford Law Review Online*, *66*(65), 65–72.

Kitchin, R. (2014). Big data, new epistemologies and paradigm shifts. *Big Data & Society*, *1*(1), 1–12.

Kramer, A. D. I., Guillory, J. E., & Hancock, J. T. (2014). Experimental evidence of massive-scale emotional contagion through social networks. *Proceedings of the National Academy of Sciences*, *111*(24), 8788–8790.

Kroening, D. (2017, 17 February). AI will save us (or at least correct our mistakes). *The Huffington Post*. Retrieved from www.huffingtonpost.co.uk/daniel-kroening/ai-will-save-us-or-at-lea_b_14790230.html.

Lin, P. (2015, 20 April). Do killer robots violate human rights? *The Atlantic*. Retrieved from www.theatlantic.com/technology/archive/2015/04/do-killer-robots-violate-human-rights/390033/.

Lin, P. (2016, 11 July). Tesla autopilot crash: Why we should worry about a single death. *IEEE Spectrum: Technology, Engineering, and Science News*. Retrieved from http://spectrum.ieee.org/cars-that-think/transportation/self-driving/tesla-autopilot-crash-why-we-should-worry-about-a-single-death.

Lin, P., Abney, K., & Bekey, G. A. (Eds.). (2011). *Robot ethics: The ethical and social implications of robotics*. Cambridge, MA: The MIT Press.

Lyon, D. (2014). Surveillance, Snowden, and big data: Capacities, consequences, critique. *Big Data & Society*, *1*(2), 1–13.

Marks, A., Bowling, B., & Keenan, C. (2015). *Automatic justice? Technology, crime and social control* (SSRN Scholarly Paper No. ID 2676154). Rochester, NY: Social Science Research Network. Retrieved from https://papers.ssrn.com/abstract=2676154.

Marr, B. (2016, 20 December). Big Data: The 6th "V" everyone should know about. *Forbes*. Retrieved from www.forbes.com/sites/bernardmarr/2016/12/20/big-data-the-6th-v-everyone-should-know-about/.

Marx, G. T. (2002). What's new about the "new surveillance"? Classifying for change and continuity. *Surveillance & Society*, *1*(1), 9–29.

Mayer-Schönberger, V., & Cukier, K. (2013). *Big data: A revolution that will transform how we live, work, and think* (1st edn). Boston: Eamon Dolan/Houghton Mifflin Harcourt.

McCulloch, J., & Wilson, D. (2015). *Pre-crime: Pre-emption, precaution and the future*. Abingdon, Oxon; New York, NY: Routledge.

Meek, A. (2015, 14 September). Data could be the real draw of the internet of things – but for whom? *Guardian*. Retrieved from www.theguardian.com/technology/2015/sep/14/data-generation-insights-internet-of-things.

Mosco, V. (2014). *To the cloud: Big data in a turbulent world*. Boulder: Routledge.

O'Hara, D., & Mason, L. R. (2012, 30 March). How bots are taking over the world. *Guardian*. Retrieved from www.theguardian.com/commentisfree/2012/mar/30/how-bots-are-taking-over-the-world?CMP=Share_AndroidApp_E-po%C5%A1ta.

Pasquale, F. (2017, 6 April). @FrankPasquale [Twitter moment]. Retrieved from https://twitter.com/FrankPasquale.

Perry, C. (2011, 18 October). You're not so anonymous. *Harvard Gazette*. Retrieved from http://news.harvard.edu/gazette/story/2011/10/youre-not-so-anonymous/.

Polajžer, K. (2016, 20 July). BTC Living Lab. Retrieved 14 January 2017, from https://prezi.com/jxdnvlloceoc/living-lab/.

Raicu, I. (2015, 6 November). Metaphors of big data. *Recode*. Retrieved from www.recode.net/2015/11/6/11620416/metaphors-of-big-data.

Raley, R. (2013). Dataveillance and countervailance. In *"Raw data" is an oxymoron* (pp. 121–146). Cambridge, MA; London, England: The MIT Press.

Ramesh, D. (2017, 1 September). What I always wanted to know about big data* (*but was afraid to ask). Retrieved from www.7wdata.be/big-data/what-i-always-wanted-to-know-about-big-data-but-was-afraid-to-ask/.

Reynolds, M. (2017, 25 February). AI learns to write its own code by stealing from other programs. *New Scientist*, (3114). Retrieved from www.newscientist.com/article/mg23331144-500-ai-learns-to-write-its-own-code-by-stealing-from-other-programs/.

Robinson, D., & Koepke, L. (2016). *Stuck in a pattern. Early evidence on "predictive policing" and civil rights*. Upturn. Retrieved from www.teamupturn.com/reports/2016/stuck-in-a-pattern.

Rosenberg, J. (2016, 5 May). Only humans, not computers, can learn or predict. *TechCrunch*. Retrieved from http://social.techcrunch.com/2016/05/05/only-humans-not-computers-can-learn-or-predict/.

Schneier, B. (2006, 9 March). Why data mining Won't stop terror. *Wired*. Retrieved from http://archive.wired.com/politics/security/commentary/securitymatters/2006/03/70357?currentPage=all.

Selingo, J. (2017, 4 November). How colleges use big data to target the students they want. *The Atlantic*. Retrieved from www.theatlantic.com/education/archive/2017/04/how-colleges-find-their-students/522516/.

Shaw, J., & Porter, S. (2015). Constructing rich false memories of committing crime. *Psychological Science*, *26*(3), 291–301.

Siegel, E. (2013). *Predictive analytics: The power to predict who will click, buy, lie, or die* (1st edn). Hoboken, NJ: Wiley.

Smith IV, J. (2015, 9 November). "Minority report" Is real – and it's really reporting minorities. *Mic*. Retrieved from https://mic.com/articles/127739/minority-reports-predictive-policing-technology-is-really-reporting-minorities.

Smith IV, J. (2016, 10 September). Crime-prediction tool PredPol amplifies racially biased policing, study shows. *Mic*. Retrieved from https://mic.com/articles/156286/crime-prediction-tool-pred-pol-only-amplifies-racially-biased-policing-study-shows.

Valicon. (2016, 27 June). Snapchat in Instagram imata v Sloveniji več uporabnikov kot Twitter. Marketing Magazin. Retrieved 3 April 2017, from www.marketingmagazin.si/novice/mmarketing/13028/snapchat-in-instagram-imata-v-sloveniji-vec-uporabnikov-kot-twitter.

Wacquant, L. (2009). *Prisons of poverty* (Expanded edn). Minneapolis: University of Minnesota Press.

Wagner, J. (2016, 9 November). Clinton's data-driven campaign relied heavily on an algorithm named Ada. What didn't she see? *Washington Post*. Retrieved from www.washingtonpost.com/news/post-politics/wp/2016/11/09/clintons-data-driven-campaign-relied-heavily-on-an-algorithm-named-ada-what-didnt-she-see/.

Watson, S. M. (2015). Data is the new "–". *DIS Magazine*. Retrieved from http://dismagazine.com/discussion/73298/sara-m-watson-metaphors-of-big-data/.

Wikipedia. (2017a, 8 April). List of countries by number of mobile phones in use. In *Wikipedia*. Retrieved from https://en.wikipedia.org/w/index.php?title=List_of_countries_by_number_of_mobile_phones_in_use&oldid=774459335.

Wikipedia. (2017b, 19 June). Technological singularity. In *Wikipedia*. Retrieved from https://en.wikipedia.org/wiki/Technological_singularity.

Wittgenstein, L. (2005). Tractatus logico-philosophicus. The Project Gutenberg ebook. Retrieved from: https://archive.org/stream/tractatuslogicop05740gut/tloph10.txt.

Yellin, T. (2016, 16 April). Most of big data is "trash" says Netflix's Todd Yellin. *BBC News*. Retrieved from www.bbc.com/news/technology-36093007.

Žižek, S. (2008). *Violence: Six sideways reflections* (1st edn). New York: Picador.

Part II

Automated social control

2 Paradoxes of privacy in an era of asymmetrical social control

Frank Pasquale

It is the best of times, and the worst of times, for criminal law enforcement. Civil liberties advocates complain of unprecedented mass surveillance. Police worry about the signals of crime "going dark," thanks to new advances in encryption. A series of legal disputes, ranging from the Snowden revelations to the Apple/FBI encryption showdown, highlights the stakes of transparency and opacity.

These debates are traditionally framed as "privacy versus law enforcement" or "civil liberties versus anti-terrorism," with familiar characters on each side of the divide. Edward Snowden himself, for example, is seen as one of the world's great advocates of the right to privacy; intelligence agency spokesmen are the public face of efforts to stop terrorism. Recently both sides have seized on "security" as an overarching rationale for what they do. Privacy advocates say that our messages can only be secure in a world that enables pervasive encryption; authorities in both police and military roles respond that national security demands a backdoor, or some way of deciphering messages.

Even more intriguingly, advocates on either side of this divide are drawn toward what often seem like contradictory stances. For instance, Snowden and his allies at *The Intercept* widely promoted the Panama Papers leaks – an effort to expose illicit financial dealings by wealthy elites – despite the fact that the leaker clearly betrayed the trust of his or her firm's clients and violated their privacy. True, the exposure was reported as having been accomplished by a former employee of the law firm Mossack Fonseca, which had papered many of the suspect deals. But what if no whistleblower had come forward, and only government action could have exposed the Panama Papers? Should we lament such an exposure as an invasion of privacy – or, if it involved decryption, an assault on the security of our information and messaging?

I do not believe so. There is a straightforward way to unravel the ostensible conundrum here for privacy advocates. We must begin by recognizing the *asymmetry* of surveillance (and, relatedly, social control), in the present age (Pasquale, 2015a). We should next review the degree to which the very wealthy have used secrecy jurisdictions to constitute an elite that is a *law unto itself*, taking advantage of a form of deterritorialized sovereignty that has yet to develop basic norms of transparency and accountability. From these two perspectives (developed, respectively, in "Tensions in the concept of surveillance" and "Recognizing the systemic risks imposed by financial crime"), exposure of the financial affairs of high net worth individuals (HNWIs) (here defined as those with beneficial ownership of more than $10 million [US] in wealth) appears as less a violation of their privacy than as freedom of information activism necessary to understand the workings of a new governance at least as influential as that of older, more familiar forms of sovereignty. When the very wealthy use campaign contributions, regulatory capture, control of think tanks, and similar tactics to exert vast influence on governments worldwide, they should be held to standards of transparency now applied to governments.

Tensions in the concept of surveillance

Surveillance occupies a contested place in contemporary discourse. Massive international and domestic intelligence apparatuses have emerged in Europe, North America, and Asia. They occasionally make spectacular mistakes – either missing threats, or stigmatizing or jailing innocent individuals. Intelligence agencies then tend to blame their failures on incomplete implementation of surveillance. They believe that, if only they had more (and more immediate) access to larger stores of data, they would prove themselves capable of detecting, deterring, and defeating more threats.

When such surveillance is demanded, some groups in society can mobilize resources to resist it, while others cannot. For example, in the US, fraud by medical *providers* has provoked a massive, big data driven, public–private effort to audit claims from doctors and hospitals (Pasquale, 2014). But private insurance companies' overbilling has barely been addressed by the Department of Health and Human Services (US Government Accountability Office, 2016). Larger and more politically powerful, the insurers can deflect harsh monitoring and penalties endured by many smaller providers.

The same applies, a fortiori, to persons. About five years ago, I was talking to a 40-year-old friend in California who made about $13,000 (US) a year. She lived at home with her disabled parents, and took care of many of their needs, while teaching about 20 hours a week in private tutoring. I learned she had no health insurance, and I urged her to look into the Medicaid program or high risk pool in her state. She said she had, but she couldn't qualify – the program counted her parents' income and assets (as well as hers) as part of a single household, so they were too rich to qualify. "I wonder if you could put in two mailboxes with two addresses, and set up a curtain between your bedroom and bathroom and the rest of the house, and say you live apart from them?" I speculated, trying to be helpful. She gave me a strange look and said "Wouldn't that be fraud?" Indeed, it almost certainly would, and new data-mining technologies would probably be quick to pick up on it.

Compare my friend's fate to that of the residents of One Hyde Park (OHP), London, one of the most luxurious buildings in the world. Apartments there can sell for over $200 million, and are bought, sold, and recombined in esoteric ways. Few may even be inhabited – it is now commonplace for the very wealthy to simply park money in real estate that they visit a few times a year, if at all. They may not only fail to reside in the apartments – they may not even directly "own" them. Of the 76 OHP apartments sold by early 2013, 64 were held in the names of corporations, most "based" in places like Liechtenstein and the Isle of Man. The directors of those corporations may or may not be the real controllers of them. Maybe they are a legitimately constituted board – maybe they are merely "nominees" set up by a service, who've contracted away their right to actually vote as they please. Nicholas Shaxson, a leading journalist of the hidden corners of global wealth, argues that "we can conclude at least two things with certainty about the tenants of One Hyde Park: they are extremely wealthy, and most of them don't want you to know who they are and how they got their money." (Shaxson, 2013).

So while the lifestyles of the poor are constantly subject to government scrutiny, a super-class can hide even massive holdings and transactions from the tax authorities (among other prying eyes). Note, too, that the machinations at One Hyde Park help drive the surveillance that denied my friend access to health care (Pasquale, 2010). Nations lose tens or even hundreds of billions of dollars each year from tax havens. The resulting budget deficits are one reason why there never seems to be enough money for social programs. So ever more surveillance is enforced on welfare beneficiaries, directly as a result of the tax deflected from the wealthy's funds.

Thus while many think tanks funded by millionaires and billionaires may say that they are fighting for a general, universalist "right to privacy," the degree of privacy enjoyed by the rich and poor may be inversely, rather than directly, related. The wealthy can use the law to keep their affairs secret; that effort starves the state of funds, leading it to investigate and monitor the less advantaged even more intensely in order to root out "benefits fraud" – or simply to keep itself funded with fines. The United States Department of Justice documented this relationship in detail in Ferguson, Missouri: thanks to declining availability of general tax revenue (which is disproportionately paid by those earning above the median income), municipalities like Ferguson resorted to "offender-funded justice." What they lacked in taxes, they made up in fines. To capture more revenue, police had to monitor a largely African-American population ever more closely, jailing citizens for minor offenses – and then forcing them to pay for their freedom.

A US criminal justice reform system has arisen to challenge these practices. However, it is in danger of being hijacked by the very wealthy interests whose parsimony helped provoke intense scrutiny of the poor in the first place. A celebrated bill in the US Congress designed to reduce criminal penalties for nonviolent offenses also includes language that would make it much harder for federal law enforcement officials to prosecute white collar crimes. And the harder these prosecutions become, the less motivation enforcers will have to even monitor and understand complex financial dealings like those behind One Hyde Park, and myriad US secrecy jurisdictions (like Delaware, Nevada, and Wyoming, which are notoriously opaque).

For average citizens, ever more of daily life is organized around the demands of what Shoshanna Zuboff calls "surveillance capitalism": intimate monitoring of our daily lives to maximize our productivity as consumers and workers. The Chief Data Scientist of a Silicon Valley firm told Zuboff:

> The goal of everything we do is to change people's actual behavior at scale. When people use our app, we can capture their behaviors, identify good and bad behaviors, and develop ways to reward the good and punish the bad. We can test how actionable our cues are for them and how profitable for us.
> (Zuboff, 2016)

Could this kind of monitoring and influence be directed at the powerful? And if it could not be universally directed, is it legitimate?

The critical question now in criminal justice reform is not a blanket reduction in the amount of scrutiny that government can direct at citizens. The question, rather, is how to use big data to promote forms of social control that can curb the baleful influence of "out of control" financial flows. This political perspective on big data and privacy helps us better understand one of the most publicized controversies in data protection law in the past decade: Apple's refusal to develop a tool to assist the FBI's effort to reveal the data in an encrypted iPhone.

Our own devices

The Snowden revelations of 2013 were a critical moment in the history of surveillance. While isolated journalists and academics had been warning about widespread private sector assistance in illegal spying activities before 2013, the concerns were a fringe issue. The private contractor Snowden's explosive breach of NSA security changed the public dialogue by exposing in-house documentation of both conscious cooperation from tech and communications giants, and easy co-optation of their networks. The deep intertwining of state and market, already obvious among carriers, became hard to ignore among Google, Apple, Facebook, Amazon, and even tech sector also-rans.

To regain user trust, Google and Apple in particular took various steps to harden their networks and devices from snooping. As Apple asserted in a brief, it "includes a setting that – if activated – automatically deletes encrypted data after ten consecutive incorrect attempts to enter the passcode" (Brief of Apple Inc., 2016). The crux of Apple's conflict with the FBI now is whether the Bureau can compel Apple, pursuant to the All Writs Act (AWA), to write software to disable this self-destruct function, and other security features.

From a purely legal perspective, the FBI's prerogative here is a very close question (Kerr, 2016). When it comes to policy, though, there is an emerging consensus: for the sake of privacy and cybersecurity in general, don't force Apple to write this software. The slippery slope to generalized, judicially mandated decryption, readily co-opted by hackers or foreign governments, is just too steep. The former head of the CIA and NSA (Michael Hayden), a commissioner of the FTC, numerous academics, and a united front of privacy activists have made this or similar arguments. If the primary evils one is fighting are US law enforcement agencies' growing disregard for Fourth Amendment values, and data breaches by shadowy hackers, this characterization of the issue is compelling.

On the other hand, then-President Barack Obama wisely warned that "You cannot take an absolutist view on this. If your argument is strong encryption no matter what, and we can and should create black boxes, that … [is] fetishizing our phones above every other value" (Obama, 2016). David Golumbia questions Apple's slippery slope salvos (Golumbia, 2016). Nathan Newman has argued that such strong encryption could be used in a number of troubling contexts:

> A tax evasion trial of top officials at the Swiss bank UBS highlighted the ways encryption stymie tax investigations, with witnesses detailing how private bankers hide sensitive client data under "Solitaire" game tabs on secret drives on encrypted laptops or code computers with emergency passwords to vaporize data on illegal activity.
>
> Just last May, when Quebec tax authorities showed up at Uber Canadian offices, engineers in Uber's San Francisco offices tried to remotely encrypt their data in Canada. Tax investigators were seeking to find out if the company was evading local sales tax rules and found that the devices were locking up as they were seeking to review the company's data after a court order.… Unbreakable encryption is … a technological kludge that will abet more vicious problems of criminal, terrorist and corporate lawbreaking than any public good it might contribute.
>
> (Newman, 2016)

Privacy activists are right to be concerned that the FBI has misused its authority in the past, and could easily do so in the future. However, the same technology could defeat their own efforts: for example, a large firm may manage to encrypt all its privacy-violating activities and evade sanctions by doing so. Consider, for instance, a corporate wellness program that used Apple Watches as its platform for data gathering and dissemination (Ajunwa, Crawford, & Schultz, in press). There is much room for mischief here, including discriminatory treatment of sick staff, made all the easier if data transfers are rendered invisible to all but those who make them (Kasperkevic, 2016; Pasquale, 2015b).

The encryption arms race

Former NSA and CIA head Michael Hayden surprised many when he weighed in against the FBI (Page, 2016). He has stated that "America is simply more secure with unbreakable end-to-end encryption." But note that by "unbreakable," he does not mean "unbreakable by anyone." He wants the NSA to request additional appropriations to hire cryptographers to hack Apple phones. Others say the NSA could hack the phone with its present capabilities (Ashbrook, 2016).

Thus Hayden's position seems to boil down to one of institutional competence. Since he sees the NSA as the "main body" in the cyber domain, he wants it to continue in an arms race of cryptographic capabilities with leading firms like Apple (Nusca, 2016). He scoffs at the civil libertarians who fought the clipper chip in the 1990s, explaining that "we got around that" with bulk collection capabilities. He seems confident that NSA can gain an edge over whatever encryption tools firms come up with. And anyone aware of the promiscuous data sharing among US agencies documented in *Exposed* (Harcourt, 2015) can foresee the next step: following the recommendations of the 9/11 Commission report, the FBI will continue to "break down silos" between its data collection and that of other agencies (The 9/11 Commission, 2004). Indeed, we now know NSA "data will be shared with other intelligence agencies like FBI without first applying any screens for privacy" (Balko, 2016).

The encryption/decryption arms race happily envisioned by Hayden also strikes me as less than useful. As Phil Rogaway warned in a recent essay "The Moral Character of Cryptographic Work," we should be free to ask:

> [What if such] computer science is not benefiting man? Technological pessimists like Jacques Ellul, Herbert Marcuse, and Lewis Mumford certainly didn't think that it was. They saw modern technology as an interlocking, out-of-control system that, instead of fulfilling human needs, engendered pointless wants and deadlier weapons.
>
> (Rogaway, 2015)

From a US-centric perspective, a never-ending war-gaming of various encryption tactics between experts in Cupertino and Fort Meade may generate a community of practice better able to withstand, say, an onslaught of Chinese cyber attacks (darkly warned about by think tanks opining on the future of war) (Ricks, 2015; Sourcewatch, n.d.). But it is hard to distinguish the inevitable global responses to such developments as anything better than a new arms race, with ever higher stakes.

A world where the FBI must go to NSA for data like that in the San Bernardino iPhone *might* be better than one where the agency can, via the AWA, compel firms to undo their own encryption. Or a ruling adverse to the FBI might accelerate interagency collaborations (both domestically and internationally) toward a more robust "Information Sharing Environment." Apple's win in #ApplevsFBI did not magically create a wall of encryption around its communications (Field, 2016). Nor may it even keep information from the FBI itself. Nor is it even clear this information *should* be kept from the FBI as a normative or legal matter.

The landscape of modern surveillance is such a complex socio-technical-legal realm that every legal victory generates technical and social pushback, every technical advance provokes legal responses, and so on. The important point here is to examine not only the vastness of the modern state surveillance apparatus, but also its ongoing, symbiotic relationship with the so-called private sector.

The seductive appeal of CEO heroes and government villains

Why, then, has the case become such a cause célèbre? Because contemporary political confrontations increasingly coalesce around hashtags, legal battles, and climactic court decisions. The clash of parties before a judge is far more entertaining and easy to follow than the byzantine bureaucratic politics that determine information flows among our myriad intelligence, law enforcement, and homeland security agencies (Priest & Arkin, 2010). Given its repeated surveillance of First Amendment-protected activity, the FBI is hard to root for (American Civil Liberties Union, 2010). Meanwhile, citizens may cheer Apple's legal position here as enthusiastically as they embrace its devices. Bernard Harcourt's book *Exposed* describes how passionate the devotion can become:

> We strap on the [Apple Watch] monitoring device voluntarily, proudly. We show it off. Surely there is no longer any need for the state to force an ankle bracelet on us anymore when we so lustfully clasp this pulsing, slick, hard object on our own wrist – the citizen's second body lusciously wrapping itself around the first.
>
> (Harcourt, 2015, p. 124)

Taking a side in the #ApplevFBI controversy can ease the cognitive dissonance attendant on clear recognition of the increasingly feudal social relations generated by both tech giants and the deep state (Lofgren, 2016; Schneier, 2012;). Trust the state to keep us free of terror, or trust the firm to keep your data secure – and then go back to tweeting, 'gram-ing, snapping, commenting as usual.

But every solution brings new problems. Consider the lessons of ankle bracelets, a "cheap on crime" solution to mass incarceration (Aviram, 2015). Yes, if widely implemented as a perfected form of house arrest, they may keep tens of thousands out of jail. But there is an all-too-easy slide from using ankle bracelets to cheapen the cost of monitoring and controlling those who might have once been imprisoned, to using them (or similar technology) as all-purpose modes of identification, monitoring, and control of entire populations, regardless of adjudicated criminality. Technological dreams and nightmares mix: how the reduction of imprisonment may only come about via digitalization that effectively renders any part of our Deleuzian "society of control" a potentially imprisoning (or informing) impediment to action (Sadowski & Pasquale, 2015).

In #ApplevsFBI, a similarly complex dilemma is playing out. Suspicious of law enforcers, and wary of ordinary methods of holding them accountable, more Americans want their communications protected technologically. The "cryptographic commons" called for by Rogaway has failed to materialize, so they place their faith in a massive corporation (Rogaway, 2015). But who else is doing the same? What are the accountability mechanisms to keep Apple itself from betraying users? Are there forms of encryption that would render any such accountability mechanisms moot? As Jürgen Geuter asks, "How do we – as globally networked individuals living in digitally connected and mutually overlaying societies – define the relationship of transnational corporations and the rules and laws we created?" (Geuter, 2016).

There are agencies in the US government that support encryption technology, and those that attack it – and some do both (Levine, 2014; Paletta, 2016). Large firms like Apple now see commercial advantage in fighting demands for decryption in the US – and readily acceding to similar demands for access in China (Pierson, 2016). So long as our "surveillance capitalism's profits derive primarily, if not entirely, from … markets for future behavior," we can expect both firms and governments to play such double games: monitoring ever more of our lives while claiming to be protecting our privacy; bashing encryption in some contexts while hardening it in others (Zuboff, 2016). Eric Schmidt will chair DOD's Defense Innovation Advisory Board, and the holding company of a firm "outraged" by DOD's NSA, and the board of a foundation claiming to offer unbiased expertise on that very controversy (Company overview, n.d.; Goldman, 2016; Snowden leaks, 2013).

Recognizing the systemic risks imposed by financial crime

Loïc Wacquant's *Punishing the Poor: The Neoliberal Government of Social Insecurity* (2009) proposes that the "hyperinflation" of the US prison population results from a change in the state's focus: from promoting economic security to promoting physical safety via a "zero tolerance" policy for even nonviolent offenses (Phillip-Fein, 2009). According to Wacquant, media and law enforcement elites team up to "erect[] a garish theater of civic morality on whose stage political elites can orchestrate the public vituperation of deviant figures … and close the legitimacy deficit they suffer when they discard the established government mission of social and economic protection." (Wacquant, 2009). Like the "security theater" lambasted by some anti-terrorism experts, the penal system explored by Wacquant is about far more than its stated purpose of keeping good citizens safe (Mueller, 2006). Rather, it becomes what Wacquant calls "autophagous," provoking a self-renewing cycle of recidivism, widening insecurity, and ever more crackdowns, by virtue of its very brutality. Like Niklas Luhmann's social theory of "autopoietic systems," which constitute and reconstitute themselves according to an inner logic that may have little to do with the overall health or welfare of society (Seidl & Schoeneborn, 2010),[1] Wacquant's theory of "neoliberal penality" warns that widespread surveillance can be unmoored from real security concerns.

If widespread surveillance is here to stay, the best way to alleviate both Wacquant's and Wolin's concerns is to make it "equal opportunity," focused not only on the "punished poor" but also on private power.[2] Charges of improper ties between corporate and governmental entities have become commonplace in American politics. They have contributed to a long-term decline in citizen trust in government. Yet at the same time as American citizens have become increasingly distrustful of public servants, the fear of terror has made them ever more reliant on the national security state. In an age of consistent fiscal discipline for leading social programs, spending on military initiatives, homeland security, and law enforcement continues to balloon (Harcourt, 2010; Perkinson, 2010; Wacquant, 2009). It is difficult – and perhaps conceptually impossible – to estimate the balance of cost and benefit in these extraordinary investments in militarization and surveillance. However, unbalanced surveillance threatens to fuel a whole new level of suspicion of government, thereby undermining the very cooperative relations it is intended to cultivate.

The Pentagon has already "simulate[d] what would happen if the world disintegrated into a series of full-fledged financial wars." (Weiner, 2010). It should not only continue to "war-game" these scenarios, but should also implement far-reaching surveillance of markets and capital flows.

Modern financial flows are not merely menacing because a computer virus could sabotage intended trades or unravel recordkeeping systems. They are also increasingly out of control and destabilizing when all components operate as designed. And if unbreakable encryption were ever fully integrated into the financial flows of the very wealthy, already cash-strapped governments might find their ability to levy taxes fatally compromised.

Fortunately, building early warning systems into the finance system will not be as difficult or costly as the vast anti-terror apparatus. Much of the groundwork has already been laid, both technically and legally. But it will require synthesis of three now-distinct policy areas: national security, cybersecurity, and finance. The cyberwar literature has already informed national security policy with longstanding cybersecurity concerns: advanced military technology can be as much a liability as an asset if it can be hacked by enemies. However, cyberwar experts have only begun to analyze the risks entailed by massive disruptions of the financial system. Financial experts understand these risks, but only a small number of them have tried to communicate them to military authorities. Like the aviation experts of the 1990s who either failed to realize or communicate the destructive potential of a weaponized plane, finance law experts have not adequately addressed the fragility of our economic order. Far from being merely the arcane concerns of a mandarin elite, technologies of finance are a cornerstone of social order. It is time to start treating them as such.

The hidden trillions

Tax secrecy is one of the most important types of financial crime. The most fundamental tool of tax secrecy is separation: between persons and their money, between corporations and the persons who control them, between beneficial and nominal controllers of wealth. In the rarefied world of the global superrich, we finally find an environment where privacy is, indeed, a purchasable commodity.[3] Certainly there are always risks of discovery, or being taken advantage of by a disreputable tax shelter broker or shady foreign bank. But for some wealthy individuals, tax havenry is one more rite of passage on the way to membership in a shadowy global elite.

Tax havenry can't be reduced to an algorithm, but it does have several well-known steps. The items below are designed to give a sense of how the stealthy spirit away funds (and spend them) in ways designed to deflect the attention of tax authorities, creditors, and journalists.

First, one needs to choose a tax haven (or two, or ten – the more secrecy one needs, the more separate entities it may be wise to set up). Dozens of sovereign entities worldwide may be happy to oblige. For those close to the Caribbean, Barbados, the Bahamas, and Bermuda all have well-developed financial services. The Cayman Islands is rumored to have more money on deposit than all the banks in New York City. European microstates like Liechtenstein are favored destinations for the wealthy in Germany, Italy, and France. Rich families in Latin America may choose Florida, Nevada, Wyoming, or Delaware: all compete for their funds, with very few questions asked (Palan, Murphy, & Chavagneux, 2010).

Most secrecy jurisdictions fall into three main categories: small European states like Switzerland, Luxembourg, and Liechtenstein; a network focused on the City of London and former British colonies; and Asian havens (Shaxson, 2011b, pp. 28–31). Secrecy jurisdictions in each of the three zones compete against one another, and globally, to attract vast sums of money. What the secrecy jurisdictions lose in terms of effective rate, they try to make up in volume – for example, charging a 100-dollar fee on tens of thousands of corporate structures, rather than a percentage tax on the income they generate or hold.

Small island nations desperate for revenue can bend the meaning of legal terms beyond all recognition. For example, the reason why a settlor of a trust is not considered the "owner" of its assets is the *independence* of the trustees from the settlor. But the Cook Islands, Nevis, and Niue, permit settlors to retain control over the trust, and allow trusts to be "revocable and of unlimited duration" (Money Laundering Threat Assessment Group, 2005, p. 49). Like Liechtenstein's mysterious "anstalts," such "trusts" appear to be little more than convenient mechanisms for secrecy.[4]

Critics of tax havens often lament the "offshore" transfers of funds from legitimate states to relatively lawless jurisdictions (Baker, 2005; Palan et al., 2010; Shaxson, 2011b). But as tax expert Lee Sheppard observes, a global tax system full of loopholes is not some bizarre fluke caused by a few renegade microstates.[5] The secrecy jurisdictions enjoy the corporate registration fees from the endless partnerships and trusts needed to make these deals work, but the home countries could do far more to stop these abuses if they committed serious political capital to the issue. Legislators in the US and UK are perfectly aware of how this system works, but have failed to adequately address its development over decades.

The UK's solicitude for foreign corporations dates back at least to a decision regarding the Egyptian Delta Land and Investment Company. The company was established in the UK in 1904, to engage in business in Egypt, but later moved its Board of Directors to Cairo (Palan, et al., 2010, p. 113). Would such a firm be liable for British tax? The answer came in 1929: no (Palan, et al., 2010, p. 126). Whatever the merits in the case of Delta Land, the precedent established a remarkable loophole: firms could establish companies in the UK, permit them to be controlled by actors located overseas, and deflect taxes in both locations (Palan et al., 2010, p. 126, quoting Picciotto, 1992). To the UK tax authorities, they'd cite *Delta Land:* the real activity of the firm was happening overseas. They would then deny the jurisdiction of overseas authorities, since they were incorporated in the UK.

Such two-faced advocacy may seem like an almost trivially simple scam, but schemes like it are highly valued by a critical mass of members of the elite tax bar. A global "alternative minimum tax" might help governments avoid the problem of stateless income, if they could adequately coordinate. Automatic information exchange between tax authorities, and a common legal identifier for companies and entities associated with particular firms, would go a long way toward shedding light on this kind of duplicity. So, too, would country-by-country accounting, instead of the current, opaque system of corporate financial reporting. But before getting into the long, and often frustrating, story of disclosure here, we need to explore a few ways in which British feudal, imperial, and colonial legacies pushed tax havenry to central importance in the global economy.

The feudal aspects of tax havenry persist at the core of England, where the City of London Corporation (CLC) operates. As journalist Nicholas Shaxson explains, over centuries, English sovereigns have needed financing to engage in wars or other projects. In exchange, the City (shorthand for London's elite finance firms) has extracted favors for the CLC that give it something approaching semi-sovereign status (Shaxson, 2013).

States within the US play a similar role. For example, the US states of Delaware, Nevada, and Wyoming each allow the formation of corporations without registration of the beneficial owners (Tax Justice Institute, 2015). A one-story building in Wilmington, Delaware is the official home of over 200,000 "businesses." (Clark, 2009). According to one expert, "In the US, [business incorporation] is completely unregulated. Somalia has slightly higher standards than Wyoming and Nevada." (Carr & Grow, 2011).

What do these "incorporations" accomplish? A range of tools create secrecy. For example, there are "bearer shares" for corporations: only their serial numbers are recorded (if that), rather than the names of the owners of the shares (Money Laundering Threat Assessment Working Group, 2005, p. 47). The beneficial owner of the company may insert "nominee" shareholders and directors. According to a money laundering report put out by several federal agencies in 2005, "nominee incorporation services (NIS) establish US shell companies and bank accounts on behalf of foreign clients" (Money Laundering Threat Assessment Working Group, 2005, p. 47).

One specialty is the *shelf* corporation – so called because it is usually a shell corporation, established years before it is sold, that is "put on the shelf" to age and to engage in a series of transactions characteristic of what an actual, bona fide firm would do. The longer a shelf corporation "ages," the pricier it is. Like a fine bottle of wine, a corporation "shelved" in the 1990s and properly tended to may sell for over $10,000 (Carr & Grow, 2011).

The proliferation of corporations can quickly give rise to mind-numbing complexity. Consider, for instance, this journalistic account of British corporation formation:

> By law, all British companies must have two things: a director, who runs the company, and a shareholder, who owns it. They can also (but do not have to) have a secretary who, in a legitimate operation, takes on much of a director's work. To create shelf companies, company formation agents need directors and secretaries. And this is when things start to get complicated, because they use other companies to do jobs designed for humans.
>
> As soon as companies were involved in owning other companies, as well as being their directors and secretaries, it became extremely difficult to discover who really controlled them (i.e. who was the "beneficial owner", the person who received any benefit from the company).
>
> (Bullough, 2016)

Why are states enabling this gutting of traditional principles of transparency and accountability? Secrecy jurisdictions have some legitimate purposes: for example, assembling plots of land for an urban development may only provoke holdouts if it is done too openly. But more ideological and narrowly pecuniary ends seem to dominate. Wyoming, for example, is one of the most conservative states in the US. Dismayed by national regulation, its leading politicians are probably pleased by its rise as a secrecy jurisdiction: activity that cannot be detected cannot be regulated.

A 2005 interagency report observed that Delaware, Nevada, and Wyoming "offer company registrations with cloaking features such as minimal information requirements and limited oversight that rival those offered by offshore financial centers" (Money Laundering Threat Assessment Working Group, 2005, p. 47). The FBI has estimated that nominee incorporation services launder tens of billions of dollars per year from Russia. Far from being a paragon of rectitude in global tax matters, the US is just one more convenient way for a global elite to shelter funds (Money Laundering Threat Assessment Working Group, 2005, p. 48).

Second, wealth tax avoiders or evaders will create a "second self" via a corporation or trust that can be established in the tax haven of their choice. To minimize detection, a person may split their investments and sources of income into myriad "related entities" crafted to minimize tax liability. The United Nations noted in a 1998 report that:

> the principal forms of abuse of secrecy have shifted from individual bank accounts to corporate bank accounts and then to trust and other corporate forms that can be purchased readily without even the modest initial and ongoing due diligence that is exercised in the banking sector.
>
> (Blum, Levi, Naylor, & Williams, 1998, p. 57)

Several entities can be used for this purpose, ranging from offshore trusts to private foundations to limited liability companies to "off the shelf" companies stored in lawyers' offices. Each has its benefits and drawbacks; for simplicity's sake, let's consider the trust for now. The creator of a trust is known as its "settlor;" he can supply funds to it. Then, he can write rules for how the "trustee(s)" (or manager[s]) of the trust can disburse or invest the funds. While there is supposed to be some independence of the trustees from the settlor, it is relatively easy for the settler to get around this rule. The settlor may, for example, hire "nominees" to serve as trustees, sworn to secrecy never to tell whose orders they actually follow.[6]

Consider, for instance, the charitable trust. A wealthy person may, for example, want to convert stock or bond holdings into a charitable donation. She could simply sell the stock, pay taxes on the capital gain involved, and give some portion of the cash to a favored charity. Alternatively, she can donate the asset to a charitable trust that she controls. A trust is simply a legal relationship where one party (a trustee) holds property for the benefit of another party (beneficiary). Remainder trusts are a small variation on the idea, where the trust may, say, pay a set return to the settlor of the trust (the person who donates assets to the trust), with the understanding that the remainder of the trust is to be transferred to the beneficiary once the trustee dies. So, for example, a donor (the settlor) might give, say, $10,000 of bonds to a trust, with the understanding that the donor is to receive all the interest payments from the bonds until her death. At that point, the bonds are the property of the beneficiary, on whatever terms the donor may have set.[7]

In his book *Perfectly Legal*, David Cay Johnston (2005, pp. 8–10) describes how a famous tax lawyer developed an "accelerated charitable remainder trust" to help wealthy families. One version of the plan, offered to Bill Gates, would have allowed him to "reap $200 million in profits on Microsoft stock without paying the $56 million of capital gains taxes that federal law required at the time." (Johnston, 2005, pp. 8–10). The plan's outline was relatively simple: rather than granting Gates some dividends each year, 80 percent of its contents would go back to him for two years. Rather than simply paying 28 percent tax to the government (and getting 72 percent of the capital gains for himself), this arrangement would give 96 percent of the gains to Gates, and 4 percent ($8 million) to a charity. The government would get nothing – in fact, less than nothing, since the deal would also create a tax deduction for Gates that would reduce his income taxes by about $2 million (Johnston, 2005, pp. 9–10).

Johnston notes that we don't know if Gates used this plan – individual income tax returns in the US, unlike in Norway, are secret (Blank, 2011). But Blattmachr's idea spread to many other wealthy individuals' tax planners. Some of the more conscientious ones alerted the Internal Revenue Service (IRS), which announced in July 1994 that it did not consider the strategy valid. But we still don't know how many wealthy individuals got away with the scheme. And the IRS's reasoning in its announcement, on why it would not recognize the "accelerated charitable remainder trust" as a means of reducing tax liability, only emboldened Blattmachr to tweak it into "son of accelerated charitable remainder trust." The IRS blocked that new scheme, too – but not before tens of millions more dollars may have slipped out of its reach (Johnston, 2005, p. 10).

Interviewed in 2012 about charitable remainder trusts, Blattmachr bragged about setting up hundreds of them in the 1990s (Drucker, 2012). He designed the "charitable" aspect of them as a "throwaway":

> The main benefit from a charitable remainder trust is the renting from your favorite charity of its exemption from taxation. I used to structure them so the value dedicated to charity was as close to zero as possible without being zero.

There are few clearer statements of the ethics of tax avoidance: twist laws on complex financial transactions in ways that assure minimum tax liability for clients; then, take some small part of the enormous tax liability avoided as a fee for services rendered (Drucker, 2012).

A third step involves creating a bank account for the trust. Banks large and small, around the world, vie for such funds, touting their reputations for "complete" or "absolute" confidentiality. Some US enforcement actions (and growing pressure from the EU) have eroded the bank secrecy laws (and norms) that created Swiss banks' reputation as discreet, "no questions asked" holders of funds (Goessl, 2013; Voreacos, Wille, & Broom, 2011). But in the world of tax havenry, authorities are playing a bit of a game of "whack a mole;" as soon as one group of entities fades in attractiveness, another tends to take its place. The Bank of Singapore has become a popular destination for funds, as are "private client" branches of other established banks (Schwartz, 2012; Tyndale, 2009, pp. 118–119).

A fourth step involves moving one's money into the bank account (created in step 3) of the trust, company, or foundation created in step 2. There are many ways to do so. The honest rich may just wire the money over, knowing that even though this creates a paper trail, there are many ways to hide the funds that may be thrown off by the trust once money has been sent overseas. More dubious measures are also available. For instance, brokers may help an American find someone who needs to get money into the United States from the tax haven they've chosen. At that point, the American and the non-American can each switch bank accounts, and no one need be the wiser. This switch violates money laundering laws, but it can be very difficult for government to detect, especially given its limited resources.

Another strategy is to initiate a fake lawsuit. The trust you establish may sue you, and you can settle out of court for an undisclosed sum pursuant to a confidentiality agreement. This is also a violation of the law, but how likely is it that overwhelmed courts are going to look behind well-papered trusts and nominee corporations? An even shadier strategy involves hiring someone to hire a courier to, say, ship or fly a huge sum of cash or gold. Again, this is almost certainly criminal, but given the lack of security in certain overseas jurisdictions (and the impossibility of the Federal Aviation Administration (FAA) or Department of

Homeland Security (DHS) screening all the contents of every private plane), it's not terribly likely to be caught, either.[8] Put enough intermediaries between the beneficial owner and the mule, and tracking the funds becomes even harder.

At the fifth stage, the offshore tax evader will usually invest the money, in instruments ranging from staid Treasury Bills to commodities to exchange traded funds. He may only report the income (if at all) to a jurisdiction with a very low (or perhaps zero) tax rate. (Operating from the registration fees they earn from tens of thousands of shell entities, many tax havens don't need to tax income booked on their soil.) The money can accumulate in the trust for years. And when it is desired for use, it might be sent to another bank account for another trust in another tax haven – starting the cycle again, repeated as needed, until an impenetrable web of shell companies prevents almost anyone from understanding the money's true origins and destinations. As a report recently noted, the "international financial system has evolved to accommodate a wide array of illicit activities, and shell companies and banking havens make it easy to camouflage transfers, payment orders, and copies of checks" (Keefe, 2013).

From tax haven to secrecy jurisdiction

Journalist Nicholas Shaxson has exhaustively covered the rise of what he calls "secrecy jurisdictions" – an important shift in nomenclature which gets at a critical flaw of current international tax systems. Shaxson argues that:

> The term "tax haven" is a bit of a misnomer, because such places aren't just about tax. What they sell is escape: from the laws, rules and taxes of jurisdictions elsewhere, usually with secrecy as their prime offering. The notion of elsewhere (hence the term "offshore") is central.
>
> (Shaxson, 2011a)

For Shaxson, the term "tax haven" mixes up two aspects of unfairness: tax rates that are too low to support the basic needs of the state, and practices of recordation that allow taxpayers to hide or misreport income. The first issue is often far more controversial than the latter, because one's views on the optimal level of taxation depend on a larger set of commitments to a certain size of state. But the latter ought to be less controversial, since whatever tax rules are set, they ought to be applied to all – not skillfully avoided by those with the lawyering (and chutzpah) to hide their wealth and earnings (Shaxson, 2011b).

Admittedly, the latter principle can be disputed too, particularly overseas. For example, many of the wealthy in authoritarian states complain bitterly about politicized application of tax laws. They want to keep their funds secret so that even in the case of a great political reversal at home, they can find financial security abroad. The tax haven "how to" literature sometimes reads as if its entire intended audience were brave dissidents fleeing cruel statists (as well as a smattering of honest businesspersons trying to shield hard-earned fortunes from rapacious exes).

If secrecy jurisdictions made a good faith effort to verify such stories, perhaps they would make a compelling argument for some kind of financial secrecy. (And even if we did want to accept such a principle, would it not make sense to limit it to some multiple of a decent estimate of living expenses – rather than, as is now the case, an excuse for hiding dynastic wealth?) However, there is precious little evidence that any of the world's 60 or so secrecy jurisdictions make an effort to separate "worthy" applications for financial privacy from unworthy ones.

Secrecy arose in such jurisdictions in order to keep higher tax countries from fully understanding their citizens' financial affairs. For example, while Swiss bankers were shielding wealth going back to the 1700s, the country's criminalization of revelations about bank accounts only happened in 1934, in response to a French scandal (Shaxson, 2011b, p. 28).[9] Not to be outdone, the Cayman Islands passed a "Confidential Relationships (Preservation) Law" that not only criminalized the revelation of financial information, but also provided for a jail term merely for asking about it.[10]

Secrecy jurisdictions rose to prominence during the 2012 presidential election, as journalists realized just how little they knew about candidate Mitt Romney's wealth. Romney had tried mightily to obscure his tax maneuvers for much of his campaign, only to finally capitulate and release certain years' returns after speculation reached a fever pitch. Romney had about $30 million in several funds in the Cayman Islands. The most notable tax trickery involved Romney's massive individual retirement account (IRA), which held at least $80 million, despite limits on contributions to such accounts. Tax experts stated that Romney probably accomplished this feat by directing his IRA to invest in a special class of Bain Stock with a temporarily depressed value (Shaxson, 2012).

Given the importance of dynasties in American politics, perhaps the aspect of Romney family tax planning with the most future importance is his shift of about $100 million in wealth to a family trust (which wasn't even counted as part of his $250 million net worth). It's quite possible that Romney's five sons will never need to work, thanks to those funds. Realizing that "gifts" could circumvent normal estate taxes, Congress restricts tax-free gifts to $13,000 per year.[11] Had Romney simply given cash to the family trust, a tax bill of at least $25 million would have been due. But he appears to have avoided that bill, too, perhaps by playing with the initial valuation of "carried interest" deposited in the trust. We do know that he used an "intentionally defective grantor trust" to shift tax bills advantageously (Dickinson, 2012).

Romney used another trust, the Charitable Remainder Unitrust (CRUT), as a tax vehicle, too. In the 1990s, had he simply sold stock for a gain, he would have paid combined federal and Massachusetts taxes of 40 percent. Instead, he gave it to a trust, which then paid him a set return, and some of the principal. Securities and Exchange Commission (SEC) filings also revealed that some portion of Romney's fortune was located in an offshore shell company in Bermuda "wholly owned by W. Mitt Romney," named Sankaty High Yield Asset Investors (Drucker, 2012; Shaxson, 2012).

By the end of campaign 2012, Romney's envelope-pushing tax strategies helped sink his presidential campaign. It was a particularly potent issue after a hidden video camera revealed that he had spoken condescendingly about the 47 percent of Americans who "paid no taxes."[12] Romney had been in politics for over a decade, knowing full well that there would be demand for information (and possible leaks about) his tax affairs. One can only wonder what other Very High Net Worth Individuals with lower profiles manage to accomplish, knowing how low the reputational stakes are for their own affairs (Klein, 2012).

The distinctiveness of the secrecy jurisdiction

There is a proper balance between maintaining privacy for tax-payers and law enforcement capacity for tax authorities. Secrecy jurisdictions upset the balance by making it essentially impossible to discover wrongdoing, or to properly allocate judgments against those who have engaged in it (Leikvang, 2012).

Consider, for instance, the use of "nominee directors." A wealthy person in the US may phone a company in a secrecy jurisdiction and ask it to establish a shell company controlled by a few directors. The directors are probably not a matter of public record – but for an extra layer of secrecy, "nominee directors" may be utilized. Contracts or less formal understandings may bind these "nominees" to do whatever the wealthy person wants them to do. The wealthy person can then transfer cash into the shell, secure in the knowledge that tax authorities at home can't trace the beneficial ownership of the funds (or any income they generate) back to their real controller and owner (Leigh, Frayman, & Ball, 2012).

In *Tax Havens Today*, Hoyt Barber reports that in Caribbean nations, these shell firms are often called "international business companies" (IBCs) (Barber, 2007, pp. 63–73). We tend to think of the corporate form as a collection of people, sometimes with thousands of shareholders and a dozen or so directors. But Caribbean IBCs generally require only one shareholder and one director – and another corporation, trust, or partnership can serve as that director. Incorporation services provide "nominee directors" who may be (on the) boards of hundreds or thousands of companies. Contact your nominee via your lawyer, and secrecy may be perfected: attorney client privilege protects your conversation with your lawyer, while the nominee is likely barred by the tax haven's law from sharing any information about his directions or doings with respect to registered trusts. To be sure, there are crime-fraud exceptions to attorney–client privilege, but in particularly murky areas of tax and corporate law, how likely is it for either the client or the attorney to *know* they are engaged in clearly criminal behavior – or for those on the outside to be able to detect what's going on? (*In re* grand jury proceedings, 1996). This mass market model almost guarantees a massive increase in sheltered money – for the tax havens are constantly undercutting one another, and need to make up for the discounts with high volume (Barber, 2007).

Even better, "board meetings" may be held by telephone (though logistical concerns certainly must be minimal when there are only one or two people involved). And if arranging such meetings before, say, a disbursement of funds is to be accomplished, "actions can be ratified after the fact" (Barber, 2007, p. 64). The purposes and functions of the corporate form are lost in the shuffle. Certain jurisdictions also make it easy for the unscrupulous to lard trusts with even more layers of secrecy. Nevis, for instance, permits settlors to retain control over the trust and allows trusts to be revocable (Barber, 2007, p. 52). Other tax havens "lack official registers that contain corporate information or that are substantially limited in the scope of information they contain, especially for companies that do not actually transact any business in the jurisdiction" (Leikvang, 2012, p. 301). No records can mean near-undetectability for savvy players.

This probably sounds dodgy to any normal taxpayer, and more than a bit risky. How can overseas "investors" be sure that the nominee directors won't run off with their money? This is where the pricy bit of tax havenry tends to come in. The average tax haven client is not going to ring up the first "offshore wealth protection" firm that comes up on a Google search. Long hours at the country club, company picnics, or business school reunions teach the discerning which contacts can minimize the chances of detection (Froomkin, 2011).

Confronting the scourge of tax havens

If widespread surveillance is here to stay, it should be "equal opportunity." "The moral decay of our society is as bad at the top as the bottom," wrote one commentator in the wake of the financial crisis, and scrutiny should be applied to those in all walks of life if it is applied to any (Oborne, 2011). Charges of improper ties between corporate and governmental entities have become commonplace in American politics. They have contributed to a long-term decline in citizen trust in government. Unbalanced surveillance threatens to fuel a whole new level of suspicion of government, thereby provoking the very disloyalty it is now aimed at monitoring and deterring. It is time to soberly reevaluate "threats" to national order, and what types of surveillance is needed. Income should be routinely monitored by the United States intelligence apparatus, and its ample surveillance capacities should be part of the implementation of the recent Foreign Account Tax Compliance Act (2010) (FATCA).

Thanks in part to a series of embarrassing investigations of tax havens by Senator Carl Levin, Congress finally took a step toward curbing the secrecy of offshore accounts in 2010. The FATCA requires foreign financial institutions (FFIs) to report financial information about accounts held by United States persons, or pay a withholding tax (Grinberg, 2012, pp. 3, 23). Congress enacted FATCA in response to the problem of international tax evasion and several high-profile court cases (Behrens, 2013, p. 213). It had become apparent that the United States lacked an effective system for preventing US taxpayers from using offshore accounts to avoid paying US taxes (Harvey, 2012, p. 474). The existing regime was inadequate because the IRS lacked the ability to identify foreign source income and determine an account's beneficial ownership (Book, 2012, p. 426). The IRS, Treasury, and congressional staff proposed and drafted

FATCA in 2009 and 2010 (Harvey, 2012, p. 476). The overall goal was to "deter and identify patterns suggestive of the use of offshore accounts to evade tax on domestic income earned by closely held businesses" (Grinberg, 2012, p. 35; Harvey, 2012, p. 487).

FATCA is designed to address each of the layers of secrecy commonly used by tax evaders. Consider, for instance, the role of shell and shelf companies, and trusts. The law requires reporting both on accounts held directly by individuals and on interests in accounts held by shell entities for the benefit of US individuals. It also covers foreign entities with significant United States ownership (26 U.S.C. §1471(c)(1)). The law also shifts the responsibility for reporting from the taxpayer to financial institutions. It hits those FFIs where it hurts: if they do not report, the law requires withholding on a wide range of payments from the United States to those same financial institutions (Grinberg, 2012, p. 24). FATCA also requires participating FFIs to withhold on payments to nonparticipating FFIs, which is supposed to create a divide between compliant and noncompliant entities (Grinberg, 2012, p. 24).

While FATCA is a step forward, it also highlights the limits of unilateral action in the tax realm. It applies to *foreign* entities, not US ones, provoking protests of hypocrisy and unfairness (Grinberg, 2012, p. 25). Some FFIs complain that compliance with FATCA may require them to violate contractual relationships with account holders, as well as data protection, bank secrecy, or other laws of the jurisdiction in which they are located or incorporated (Grinberg, 2012, p. 25; Nelson, 2012, p. 389). One skeptical commentator has concluded that "FATCA almost certainly cannot solve the problem of US taxpayers' offshore accounts without the cooperation of non-US governments" (Morse, 2012, p. 530; see also Nelson, 2012, p. 423). A former Senior Advisor to the IRS Commissioner, who was significantly involved in the development of FATCA, argues that "the ultimate long-term success of FATCA may depend upon whether other countries adopt some version of FATCA, or at least adopt detailed customer due diligence procedures of the type embedded in FATCA" (Harvey, 2012, p. 488).

Some nations are responding positively. France, Germany, Italy, Spain, and the United Kingdom have expressed approval of FATCA and stated their intentions of enacting similar laws (Berger, 2013, p. 75). Gradual implementation of the law eases their position somewhat. These developments should give some hope for FATCA's ability to reduce the problems of tax evasion and avoidance. Of course, even if foreign financial institutions report both US and foreign source income for US taxpayers, determine if U.S. taxpayers are the beneficial owners of foreign shell entities, and review all customer accounts within the affiliated group to identify US taxpayers, the IRS will still need to enforce the law (Harvey, 2012, p. 480). Strapped for resources by a hostile Republican majority in the House of Representatives, the agency has become increasingly skittish about going after well-financed tax evaders. Nevertheless, a "global regime to improve cross-border administrative assistance" should be a bipartisan goal of all those committed to the rule of law (Grinberg, 2012, p. 38).

Without coordinated international action, we can only expect more wealthy taxpayers to opt out of a tax system that grows ever more arbitrary and unfair. Out-of-control financial transactions are leaving states not merely powerless to tax massive money flows; states now profess that they can barely discern how much money their citizens earn, save, and invest. Given the American anti-terror surveillance apparatus surveyed in the third part of this book, that claim is hard to credit on its face. One recent study notes that "forming an anonymous shell company is as easy as ever … demonstrat[ing] that we are much too far from a world that is safe from terror" (Baradaran, Findley, Nielson, & Sharman, 2014). The homeland security apparatus has successfully lobbied for extraordinary funding for tanks for rural towns, monitoring of Occupy activists, and a surfeit of flat screen televisions for fusion centers (Clark, 2012). But when it comes to striking at the heart of terror *financing* – a critical aspect of nearly any highly destructive operation – the prerogatives of financiers and the secretive wealthy trump homeland security concerns. The wealth defense industry, devoted to exploiting secrecy jurisdictions and tax havens for its dubious clients, remains as strong as ever, despite growing knowledge of just how destructive its methods can be.

The migration of monetary recordkeeping to internet-enabled computer databases can either retard or enhance the ability the dishonest wealthy to hide income from legitimate tax authorities (Krantz, 2010). In a world of unmonitored, totally encrypted capital flows, concentrated wealth could purchase power in a way that fundamentally threatens the state's ability to finance itself.[13] However, smart monitoring of electronic data flows could also help states avoid that scenario – and the varied lesser challenges to state authority that lead, on average, to hundreds of billions of dollars of illicit financial flows each year, and tens of billions more in lost tax revenue (Avril, 2009; Baker, 2005). The question is whether we begin to rationalize the threat assessments of the intelligence apparatus to include financial crimes – or continue to ignore this clear and present danger to social order.

William Scheuerman (2004) has underscored the importance of rapid and flexible administration in his work on the "social acceleration of time." Scheuerman cites a succession of social theorists who call ours a "distinctly high-speed society," (Scheuerman, 2004, p. 4, citing "Zygmunt Bauman, Manuel Castells, Anthony Giddens, David Harvey, and Reinhardt Koselleck"; Wolin, 1998). Bush Administration officials have used the rhetoric of speed to justify extraordinary departures from past law enforcement practices (*Wartime Executive Power*, 2004, p. 15). As Andrew Bacevich explains, such thinking has also deeply influenced the military establishment:

> Numerically large armies … were now problematic. Small contingents of highly trained, high tech ground forces moving like quicksilver: Here was the template for all future US military operations.… Swift, precise, flexible, agile, adaptable: These qualities had now become hallmarks of US military operations.
>
> (Bacevich, 2010, p. 174)

Translating this type of flexibility and agility into the executive branch depends on recognition of the urgency of the threat that large financial institutions pose. President George W. Bush, for instance, explicitly introduced several of his initiatives by reminding the nation that we are "at war," and that his authority had commensurately expanded in order to address the extraordinary problems this situation had created. A nation constantly reminded of the fickleness of bond buyers, the specter of hyperinflation or deflation, the austerity induced by tax evasion and avoidance, and the disastrous consequences of a rapid sell-off of foreign-owned US government debt, might want to consider whether it is also entering a long period of financial emergency that justifies intensive intelligence-gathering focused on its markets.

State actors must pursue this agenda before public suspicion of interlocking state-corporate elites further erodes the legitimacy of the state. Since the late 1990s, a litany of tax and accounting scandals have devastated trust in both regulated entities and the regulators who are supposed to assure a fair playing field for business competition. The public may be becoming inured to this state of affairs, as stories of epic malfeasance for battle (and often fail to obtain) prominence in newsfeeds peppered with gossip, sports, and personalized "news." In an era of a fragmented public sphere, it is futile to expect consumer or voter mobilization against any single firm's sharp or illegal practices – especially when a parade of corporations is lambasted for wrongdoing on a weekly basis. Rather, present state authorities must assert the right to monitor, tax, and regulate firms pursuant to extant legal authorities.

Notes

1 Teubner extends Luhmann's work in Scheuerman, last chapter (Scheuerman, 2004, pp. 210–218, citing Teubner [1983]).
2 Larry Cata Backer's (2004) work on Sarbanes-Oxley as a surveillance mechanism is one good illustration of policies that need to be extended to real-time monitoring. Jaron Lanier's project of "formalized financial expression" would aid in this process (2010, p. 112). See also Wolin (2010).
3 Firms sell reputation management services. Setting up nominee directors and shell companies is another way to assure that negative information about oneself is not widely known.
4 For a description of anstalts, see Palan et al. (2010, pp. 94, 117); see also Storm survivors (2013).
5 "[N]one of the regulatory arbitrage with the banking laws, with the insurance laws, with the securities laws … is possible without the tolerance of the home governments of these … companies" (Understanding international tax havens, 2013).
6 For more on trusts generally, see Tritt (2011). For more on the "nominee trust," see *In re* Grand Jury Subpoena (1992).
7 For instance, could all be distributed at once, or paid over a bit at a time. For more on charitable trusts, see Eisenstein (2003).
8 For more on abusive tax strategies involving trusts, see Internal Revenue Service (n.d.).
9 See Palan et al. (2010, pp. 120–121), debunking myth that Switzerland in 1934 acted to protect the assets of Jewish individuals.
10 In the Cayman Islands, "[y]ou can be jailed for up to four years just for *asking* about such information" (Shaxson, 2012).

11 The limits are per-donor, per-donee.
12 This was wrong, too: Romney should have said "federal income tax." State and local taxes, often regressive, are paid by nearly everyone in the US (Hamill, 2008). Moreover, payroll taxes for Medicare and Social Security are paid by all workers.
13 Max Weber (1918) defined state authority as a "monopoly on the legitimate use of force."

References

Ajunwa, I., Crawford, K., & Schultz, J. (in press). Limitless worker surveillance. *California Law Review*. Retrieved from http://papers.ssrn.com/sol3/papers.cfm?abstract_id=2746211.

American Civil Liberties Union. (2010, August 11). *Policing free speech*. Retrieved from www.aclu.org/files/assets/policingfreespeech_20100806.pdf.

Ashbrook, T. (2016, March 1). Apple, the FBI, and your privacy [Radio program episode]. In *NPR: On Point*. Retrieved from http://onpoint.wbur.org/2016/03/01/apple-fbi-san-bernardino-encryption.

Aviram, H. (2015). *Cheap on crime: Recession-era politics and the transformation of American punishment*. Oakland, CA: University of California Press.

Avril, H. (2009, September 16). Q & A: Political elites ensure continuing flight of dirty money. *Inter Press Service News*. Retrieved from www.ipsnews.net/2009/09/qa-quotpolitical-elites-ensure-continuing-flight-of-dirty-moneyquot/.

Bacevich, A. J. (2010). *Washington rules: America's path to permanent war*. NY: Metropolitan Books.

Backer, L. C. (2004). Surveillance and control: Privatizing and nationalizing corporate monitoring after Sarbanes-Oxley. *Michigan State Law Review, 2004*, 327–440.

Baker, R. W. (2005). *Capitalism's achilles heel: Dirty money and how to renew the free-market system*. Hoboken, NJ: John Wiley & Sons, Inc.

Balko, R. (2016, March 10). Surprise! NSA data will soon be routinely used for domestic policing that has nothing to do with terror. *Washington Post*. www.washingtonpost.com/news/the-watch/wp/2016/03/10/surprise-nsa-data-will-soon-routinely-be-used-for-domestic-policing-that-has-nothing-to-do-with-terrorism/.

Baradaran, S., Findley, M., Nielson, D., & Sharman, J. (2014). Funding terror. *University of Pennsylvania Law Review, 162*, 477–536.

Barber, H. (2007). *Tax havens today: The benefits and pitfalls of investing offshore*. Hoboken, NJ: John Wiley & Sons.

Behrens, F. (2013). Using a sledgehammer to crack a nut: Why FATCA will not stand. *Wisconsin Law Review, 2013*, 205–236.

Berger, M. A. (2013). Not so safe haven: Reducing tax evasion by regulating correspondent banks operating in the United States. *Journal of International Business and Law, 12*, 51–88.

Blank, J. D. (2011). In defense of individual tax privacy. *Emory Law Journal, 61*, 265–348.

Blum, J. A., Levi, M., Naylor, R. T., & Williams, P. (1998). *Financial havens, banking secrecy and money-laundering*. New York: United Nations.

Book, L. (2012). Offshore accounts, Corporate income shifting, and executive compensation. *Villanova Law Review, 57*, 421–432.

Brief of Apple Inc. in support of motion (2016, February 25). *In re* search of an Apple iPhone seized during the execution of a search warrant on a black Lexus IS300,

California license plate 35KGD203. Retrieved from https://assets.documentcloud.org/documents/2722199/5-15-MJ-00451-SP-USA-v-Black-Lexus-IS300.pdf.

Bullough, O. (2016, April 19). Offshore in central London: the curious case of 29 Harley Street. *Guardian*. Retrieved from www.theguardian.com/business/2016/apr/19/offshore-central-london-curious-case-29-harley-street.

Carr, K., & Grow, B. (2011, June 28). Special report: A little house of secrets on the Great Plains. *Reuters*, Retrieved from www.reuters.com/article/oukwd-uk-usa-shell-companies-idAFTRE75R22L20110628.

Clark, A. (2009, April 10). Welcome to tax-dodge city, USA. *Guardian*. Retrieved from www.guardian.co.uk/business/2009/apr/10/tax-havens-blacklist-us-delaware.

Clark, C. S. (2012, October 2). Homeland Security's fusion centers lambasted in Senate report. *Government Executive*. Retrieved from www.govexec.com/defense/2012/10/homeland-securitys-fusion-centers-lambasted-senate-report/58535/.

Company Overview of New America Foundation. (n.d.) *Bloomberg*, www.bloomberg.com/research/stocks/private/board.asp?privcapId=49437705.

Dickinson, T. (2012, October 12). Mitt Romney's tax dodge. *Rolling Stone*. Retrieved from www.rollingstone.com/politics/news/mitt-romney-s-tax-dodge-20121012.

Drucker, J. (2012, October 29). Romney avoids taxes via loophole cutting Mormon donations. *Bloomberg*. Retrieved from www.bloomberg.com/news/2012-10-29/romney-avoids-taxes-via-loophole-cutting-mormon-donations.html.

Eisenstein, I. H. (2003). Keeping charity in charitable trust law: The Barnes Foundation and the case for consideration of public interest in administration of charitable trusts. *University of Pennsylvania Law Review*, *151*, 1747–1786.

Field, E. (2016, March 10). FTC official slams gov't magical thinking in iPhone row. *Law 360*. Retrieved from www.law360.com/articles/770115/ftc-official-slams-gov-t-magical-thinking-in-iphone-row.

Foreign Account Tax Compliance Act (FATCA), Title V of the Hiring Incentives to Restore Employment Act, Pub. L. No. 111–147, §§501562, 124 Stat. 71, 97 (2010) (codified at 26 U.S.C. §§1471–1474).

Froomkin, D. (2011, December 13). "Wealth defense industry" protects 1% from the rabble and its taxes. *Huffington Post*, www.huffingtonpost.com/dan-froomkin/wealth-defense-industry-p_b_1145825.html.

Geuter, J. (2016, February 18). Liberty, an iPhone, and the refusal to think politically. *Boundary2*. Retrieved from www.boundary2.org/2016/02/liberty-an-iphone-and-the-refusal-to-think-politically/.

Goessl, L. (2013, May 30). Swiss government weakens bank secrecy to give US officials info. *Digital Journal*. Retrieved from http://digitaljournal.com/article/351155.

Goldman, D. (2016, March 2). Eric Schmidt gets a job at the Pentagon. *CNN Money*. Retrieved from http://money.cnn.com/2016/03/02/technology/eric-schmidt-pentagon/.

Golumbia, D. (2016, February 25). Are "backdoors" real or virtual? the logical flaw in #AppleVsFBI. *uncomputing*. Retrieved from www.uncomputing.org/?p=1708.

Grinberg, I. (January 27, 2012). The battle over taxing offshore accounts. *UCLA Law Review*, *60*, 304–383.

Hamill, S. P. (2007). *As certain as death: A fifty-state survey of state and local tax laws*. Durham, NC: Carolina Academic Press.

Harcourt, B. E. (2010). Neoliberal penality: A brief genealogy. *Theoretical Criminology*, *14*(1), 74–92.

Harcourt, B. E. (2015). *Exposed: Desire and disobedience in the digital age*. Cambridge, MA: Harvard University Press.

Harvey, Jr., J. R. (2012). Offshore accounts: Insider's summary of FATCA and its potential future. *Villanova Law Review*, *57*, 471–498

In re Grand Jury Proceedings, 87 F.3d 377 (9th Cir. 1996).

In re Grand Jury Subpoena, 973 F.2d 45 (1st Cir. 1992).

Internal Revenue Service (n.d.). *Abusive tax evasion schemes*. Retrieved from www.irs.gov/Businesses/Small-Businesses-&-Self-Employed/Abusive-Trust-Tax-Evasion-Schemes.

Johnston, D. C. (2005). *Perfectly legal*. New York, NY: Portfolio Trade.

Kasperkevic, J. (2016, February 29). Wellness programs at work: Could your boss be spying on your health? *Guardian*. Retrieved from www.theguardian.com/business/2016/feb/29/wellness-programs-boss-spying-on-your-health.

Keefe, P. R. (2013, July 8). Buried secrets. *The New Yorker*. Retrieved from www.newyorker.com/magazine/2013/07/08/buried-secrets.

Kerr, O. (2016, February 19). Preliminary thoughts on the Apple iPhone order in the San Bernardino case: Part 2, the All Writs Act, *Washington Post: Volokh Conspiracy*. Retrieved from www.washingtonpost.com/news/volokh-conspiracy/wp/2016/02/19/preliminary-thoughts-on-the-apple-iphone-order-in-the-san-bernardino-case-part-2-the-all-writs-act/.

Klein, E. (2012, September 18). Mitt Romney vs the 47% – and himself. *Washington Post Wonkblog*. Retrieved from www.washingtonpost.com/blogs/wonkblog/wp/2012/09/18/wonkbook-mitt-romney-vs-the-47-and-himself/.

Krantz, M. (2010, July 9). Computerized stock trading leaves investors vulnerable, *USA Today*. Retrieved from www.usatoday.com/money/markets/2010-07-09-wallstreetmachine08_CV_N.htm.

Lanier, J. (2010). *You are not a gadget*. New York: Alfred A. Knopf.

Leigh, D., Frayman, H., & Ball, J. (2012, November 25). Front men disguise the offshore game's real players. *International Consortium of Investigative Journalists*. Retrieved from www.icij.org/front-men-disguise-offshore-players.

Leikvang, H. (2012). Piercing the veil of secrecy: Securing effective exchange of information to remedy the harmful effects of tax havens. *Vanderbilt Journal of Transnational Law*, *45*, 293–342.

Levine, Y. (2014, July 16). Almost everyone involved in developing Tor was (or is) funded by the US government. *Pando*. Retrieved from https://pando.com/2014/07/16/tor-spooks/.

Lofgren, M. (2016). *The deep state: The fall of the constitution and the rise of the shadow government*. New York: Viking.

Money Laundering Threat Assessment Working Group. (2005, December). US Money Laundering Threat Assessment. Retrieved from http://permanent.access.gpo.gov/lps70930/mlta.pdf.

Morse, S. C. (2012). Ask for help, Uncle Sam: The future of global tax reporting. *Villanova Law Review*, *57*, 529–550.

Mueller, J. (2006). *Overblown: How politicians and the terrorism industry inflate national security threats, and why we believe them*. New York: Free Press.

Nelson, P. (2012). Conflicts of interest: Resolving legal barriers to the implementation of the Foreign Account Tax Compliance Act. *Virginia Tax Review*, *32*, 387–424.

Newman, N. (2016, February 26). Social justice – and privacy case – for unlocking the San Bernardino iPhone. *Huffington Post*. Retrieved from www.huffingtonpost.com/nathan-newman/the-social-justice-and-privacy_b_9328118.html.

The 9/11 Commission. (2004). *The 9/11 Commission Report*. Retrieved from http://govinfo.library.unt.edu/911/report/911Report.pdf.

Nusca, A. (2016, February 19). Former NSA director: in Apple vs FBI, many top gov't officials side with Apple. *Fortune*. Retrieved from http://fortune.com/2016/02/19/hayden-apple-fbi/.

Obama, B. (2016, March 11). Remarks at South by Southwest Interactive. Retrieved from www.whitehouse.gov/the-press-office/2016/03/14/remarks-president-south-southwest-interactive.

Oborne, P. (2011, August 11). The moral decay of British society is as bad at the top as the bottom. Open Democracy. Retrieved from www.opendemocracy.net/ourkingdom/peter-oborne/moral-decay-of-british-society-is-as-bad-at-top-as-bottom.

Page, S. (2016, February 21). Ex-NSA chief backs Apple on iPhone "back doors." *USA Today*. Retrieved from www.usatoday.com/story/news/2016/02/21/ex-nsa-chief-backs-apple-iphone-back-doors/80660024/.

Palan, R., Murphy, R., & Chavagneux, C. (2010). *Tax havens: How globalization really works*. Ithaca, NY: Cornell University Press.

Paletta, D. (2016, February 22). How the US fights encryption and also helps develop it. *Wall Street Journal*. Retrieved from www.wsj.com/articles/how-the-u-s-fights-encryptionand-also-helps-develop-it-1456109096.

Pasquale, F. (2010). Access to medicine in an era of fractal inequality. *Annals of Health Law*, *19*(269).

Pasquale, F. (2014). Private certifiers and deputies in American health care. *North Carolina Law Review*, *92*, 1661–1692.

Pasquale, F. (2015a). *The black box society: The secret algorithms that control money and information*. Cambridge: Harvard University Press.

Pasquale, F. (2015b, September 21). The other big brother. *The Atlantic*. Retrieved from www.theatlantic.com/business/archive/2015/09/corporate-surveillance-activists/406201/.

Perkinson, R. (2010). *Texas tough*. New York: Metropolitan Books.

Phillips-Fein, K. (2009, September/October/November). Imprisoner's dilemma, *Book Forum*. Retrieved from http://bookforum.com/inprint/016_03/4332.

Picciotto, S. 1992. *International Business Taxation*. Westport, CT: Quorum Books.

Pierson, D. (2016, February 26). While it defies US government, Apple abides by China's orders, and reaps big rewards. *Los Angeles Times*. Retrieved from www.latimes.com/business/technology/la-fi-apple-china-20160226-story.html.

Priest, D., & Arkin, W. M. (2010). Top secret America. *Washington Post*. Retrieved from http://projects.washingtonpost.com/top-secret-america/.

Ricks, T. (2015, June 11). Singer and Coles's "Ghost Fleet": Every Army officer should read it and it's fun. *Foreign Policy*. Retrieved from http://foreignpolicy.com/2015/06/11/singer-and-coles-ghost-fleet-every-army-officer-should-read-it-and-its-fun/.

Rogaway, P. (2015, December). *The moral character of cryptographic work*. Unpublished manuscript. Retrieved from http://web.cs.ucdavis.edu/~rogaway/papers/moral-fn.pdf.

Sadowski, J., & Pasquale, F. (2015, July 6). The spectrum of control: a social theory of the smart city. *First Monday*, *20*(7). Retrieved from http://firstmonday.org/ojs/index.php/fm/article/view/5903/4660.

Schapiro, M. L. (2010, May 26). *Opening statement at the SEC Open Meeting – Consolidated audit trail*. Retrieved from www.sec.gov/news/speech/2010/spch052610mls-audit.htm.

Scheuerman, W. E. (2004). *Liberal democracy and the social acceleration of time*. Baltimore, MD: Johns Hopkins University Press.

Schneier, B. (2012, December 3). Feudal security. *Schneier on Security*. Retrieved from www.schneier.com/blog/archives/2012/12/feudal_sec.html.

Schwartz, N. D. (2012, March 11). Got $100,000? Have a cookie: Banks try luring the top 10%. *New York Times*, p. A1.

Seidl, D., & Schoeneborn, D. (2010, February 1). *Niklas Luhmann's autopoietic theory of organisations: contributions, limitations, and future prospects*. University of Zurich, Institute of Organization and Administrative Science. IOU Working Paper No. 105. http://papers.ssrn.com/sol3/papers.cfm?abstract_id=1552847.

Shaxson, N. (2011a, February 24). The tax haven in the heart of Britain. *New Statesman*. Retrieved from www.newstatesman.com/economy/2011/02/london-corporation-city.

Shaxson, N. (2011b). *Treasure islands: Tax havens and the men who stole the world*. New York: Vintage Books.

Shaxson, N. (2012, August). Where the money lives. *Vanity Fair*. Retrieved from www.vanityfair.com/news/politics/2012/08/investigating-mitt-romney-offshore-accounts.

Shaxson, N. (2013, April). A tale of two Londons. *Vanity Fair*. www.vanityfair.com/society/2013/04/mysterious-residents-one-hyde-park-london.

Snowden leaks: Google "outraged" at alleged NSA hacking. (2013, October 31) *BBC News*. Retrieved from www.bbc.com/news/world-us-canada-24751821.

SourceWatch. (n.d.). New America Foundation. Retrieved from www.sourcewatch.org/index.php/New_America_Foundation.

Storm survivors: Special report on offshore finance. (2013, February 16). *The Economist*. Retrieved from www.economist.com/sites/default/files/20130216_offshore_finance.pdf.

Tax Justice Institute. (2015, September 23). Financial secrecy index: Narrative report on USA. Retrieved from www.financialsecrecyindex.com/PDF/USA.pdf.

Teubner, G. (1983). Substantive and reflexive elements in modern law. *Law and Society Review*, *17*, 239–286.

Tritt, L. (2011). The limitations of an economic agency cost theory of trust law. *Cardozo Law Review*, *32*, 2579–2640.

Tyndale, W. (2009). *Fundamentals of offshore banking*. Vancouver, WA: Pratzen.

Understanding International Tax Havens. (2013, March 27). *NPR: The Diane Rehm Show*. Retrieved from http://thedianerehmshow.org/shows/2013-03-27/understanding-international-tax-havens.

US Government Accountability Office. (2016, April). *Fundamental improvements needed in CMS's effort to recover substantial amounts of improper payments* (No. GAO-16–76). Retrieved from www.gao.gov/assets/680/676441.pdf.

US Securities and Exchange Commission (2008, February). *In brief: FY 2009 congressional justification*. Retrieved from www.sec.gov/about/secfy09congbudgjust.pdf.

Voreacos, D., Wille, K., & Broom, G. (2011, October 24). Swiss banks said ready to pay billions, disclose customer names. *Bloomberg*. Retrieved from www.bloomberg.com/news/2011-10-24/swiss-banks-said-ready-to-pay-billions-disclose-customer-names.html.

Wacquant, L. (2009). *Punishing the poor: The neoliberal government of social insecurity*. Durham, NC: Duke University Press.

Wartime Executive Power and the National Security Agency's Surveillance Authority: Hearing Before the S. Comm. on the Judiciary, 109th Cong. 15 (2006) (statement of Alberto Gonzales, US Attorney General).

Weber, M. (1918). Politics as a vocation, speech at Munich University. In H. H. Gerth, & C. W. Mills (Eds. and trans.), *From Max Weber: Essays in sociology* (pp. 77–128). Oxford University Press: New York, 1958.

Weber, M. (1925). *Max Weber on law in economy and society*. (E. Shils, & M. Rheinstein, Trans.). Cambridge, MA: Harvard University Press, 1954.

Weiner, E. J. (2010). *The shadow market*. New York: Scribner.

Wolin, S. (1997). What time is it? *Theory and Event* 1, 1. doi:10.1353/tae.1991.0003.

Wolin, S. (2008). *Democracy incorporated: Managed democracy and the specter of inverted totalitarianism*. Princeton, NJ: Princeton University Press.

Zuboff, S. (2016, May 3). The secrets of surveillance capitalism. *Frankfurter Allegemeine*. Retrieved from www.faz.net/-gsf-8eaf4.

Zuckerman, G. (2008, January 15). Trader made billions on subprime. *Wall Street Journal*, p. A1.

3 Big data

Big ignorance

Renata Salecl

In today's society, it is not only the case that people are controlled by others (i.e. that their moves are recorded by video cameras, and data related to them are collected at every point in their lives); increasingly, people are monitoring themselves and are knowingly or unknowingly allowing various enterprises to collect their data. Although people often "sign" informed consent agreements when they use self-monitoring apps or when they engage with the Internet of things and control their environment from afar, they often ignore the fact that they are allowing corporations and state surveillance apparatuses to use their data in ways that go against their interests. With the vast new knowledge that we are dealing with in these times of big data, there is a concurrent increase in the ignorance pertaining thereto.

This chapter will first analyse the psychological mechanisms that are behind our passion for self-monitoring. Second, it will look at the way corporations exploit these passions. And third, it will address the question of why people so easily ignore the fact that data about their lives is collected, which can often be used to their disadvantage.

Self-surveillance in the era of big data

The market is flooded with devices that are supposed to help us navigate our daily lives so that we will become more productive, better organised, fitter, healthier, slimmer, and even less stressed. The expectation is that these devices will help the individual alter his or her life in such a way that it comes closer to the ideals of success and self-fulfilment. Many of the applications that we install on our smart phones rely on the idea that we will achieve these goals with the help of measurements. We can thus measure our calories, walking, running, heartbeat, menstrual cycle, and – during pregnancy – even the heartbeat of our unborn child.

If we think about previous generations, we cannot say that there was a culture of counting how many steps a person walked per day, how many calories he or she consumed, how many hours per day one was asleep, or how often one meditated per week. Sociologists researching the way post-industrial capitalism insists on an increase in productivity link the obsession with measurement to new forms of social control. The subject is constantly under pressure to produce more, to be quicker, and is especially anxious about his or her employment. Keeping track of one's productivity at the workplace has, however, in recent decades expanded into keeping track of one's private life. The ideas of achievement, success, and happiness that have been part of the dominant ideology in post-industrial capitalism have opened up the doors to the wellness industry (Cederström & Spicer, 2015) and self-help enterprises, which have become the prime promoters of the idea that with the help of proper measurement, tracking, and self-control, the subject will be able to come closer to attaining these ideals.

The first problem with plans for self-improvement is that most of the time people do not follow them for a long time; the second is that these plans often increase a person's anxiety and feelings of guilt; and the third is that the new technologies that we now use to monitor our progress allow for the collection of data about us that can be used and abused in ways we cannot easily imagine.

Embarking on a quest to change one's habits might mean constantly failing to follow a particular plan. Personal measurement and tracking appear to be strategies that can make the process of self-transformation more manageable and predictable. The numbers that we record on our devices are also supposed to help us not succumb to temptation. They seem to be contemporary self-binding mechanisms.

Looking far back into the past, Homer was aware of the necessity of self-binding, which is why his Odysseus ties himself to the mast in order to not succumb to the Siren's song. John Elster (Elster, 2016) links self-binding to various strategies whereby people try to pursue their quest to change a particular behaviour. If we, for example, want to stop smoking, we might tell everyone around us of our intention and by doing so we might be less inclined to light a cigarette in their presence. Such strategies of self-binding rely on the feelings of guilt and embarrassment that people experience in the presence of other people.

Paradoxically, the Internet allows for the creation of self-binding strategies that also rely on these feelings, even though people do not necessarily have face-to-face contact with people online. People who try to lose weight and log their food intake into an online forum daily might experience feelings of guilt when they do not follow their diet plan and, for example, admit their food indulgences to anonymous strangers online.

One cannot deny that feelings of guilt can be a powerful motivator when people try to change their habits. If online communication with anonymous interlocutors can contribute to these feelings, the question remains whether that happens also when people try to change their lives with the help of various monitoring devices.

The failure of self-monitoring

Although people fervently download apps that are supposed to monitor their progress, the majority soon forget about them, and for one reason or another stop measuring their progress. Researchers who study motivation and attempt to ascertain why apps are so easily forgotten rediscovered Aristotle's term "*akrasia*" which in antiquity described how a person acts against his or her better judgement. Today this term is supposed to depict a form of procrastination when people do not follow through with their plans (Clear, 2016).

A number of interesting studies about the failure to follow our plans with the help of tracking devices have been carried out in the field of medicine. One study looked at the link between physical activity and monetary compensation. People who were asked to monitor their physical activity and got paid for increasing the total number of steps they walked per day usually abandoned their fitness goals when they stopped receiving money for their efforts (Finkelstein, Haaland, Bilger, Sahasranaman, Sloan, Nang, & Evenson, 2016). During the study, when the subjects were financially compensated for being more physically active, it looked like they were easily able to change their lifestyle and improve their health. Although the expectation was that their increased well-being would help them continue with the plan when money ceased to be the motivating factor, for the majority of participants this was not the case. When the financial benefit ended, most of the participants became less physically active.

While it might be debatable whether money should be used as an incentive to change one's habits (Sandel, 2013), for our argument here it is interesting to look at the failure of self-monitoring through the lens of psychology and psychoanalysis. Since the late 1990s, psychological studies on willpower have relied heavily on a study (Baumeister, Bratslavsky, Muraven, & Tice, 1998) that tested the willpower of people by means of two different exercises. Baumeister and his colleagues first examined people's willpower by instructing two groups of people on what to eat. Both groups had chocolate cookies and a bowl of radishes presented in front of them. One group was asked to eat only radishes, while the other was allowed to eat cookies. The idea was to measure how much self-discipline it would take for the radish-eating group to resist the cookies. After this experiment, both groups were asked to solve puzzles which, however, were unsolvable. The surprising result was that the group that was allowed to eat cookies spent much longer on trying to solve these puzzles, while the radish eaters gave up more quickly. The explanation for this behaviour was that willpower is like a muscle that can be strengthened with regular exercise, but using it too much can deplete its strength. If we use willpower for one task we might not be as effective in using it for another. The radish eaters used up their willpower and that prevented them from being more persistent with regard to solving the puzzle, while the cookie eaters, who did not need to use their willpower in the first experiment, were able to use their willpower in the second experiment.

The failure to replicate Baumeister's experiment has led psychologists to conclude that "willpower *isn't* a limited resource, but believing that it is makes you less likely to follow through on your plans" (Burkeman, 2017). If we presuppose that trying to restrain ourselves with regard to one's temptation will exhaust our willpower and as a result we will be less likely to follow through with another project, it will actually happen that we will use less willpower in the second case. However, if we do not presuppose that there is something like "willpower fatigue", that will not happen.

Other studies in the domain of willpower have tried to tie the ability to follow self-formulated plans to change one's behaviour with emotions. Some self-help books thus advise people to observe which emotions they experience when they try to follow particular plans to change and advise people on how not to use up all their energy to deal with these emotions but rather engage in altering their environment so that it helps them pursue their goals (McGonigal, 2013).

And here we come to apps and wearable technology, which is supposed to be something that manipulates the environment in such a way that it is easier for people to follow through with plans for personal change. Since people might download many apps and buy wearable devices, but can easily "forget" to track whatever they planned to track and thus do not achieve their goals, more and more of these devices attempt to increase feelings of guilt and anxiety. The idea is that people will be more inclined to follow through with their plans if they are anxious that they will be punished for their failures.

Apps and self-punishment

A wristband device called Pavlok offers people a way to impose self-punishment when they do not follow through with their plans. A Pavlok wearer has the possibility to zap him- or herself if he or she is tempted to pursue a behaviour that he or she would like to alter. With this action, Pavlok is supposed to arouse our inner voice, which will say to us: "Wake up sleepy head … it's time to go to the gym!", "Put down those chips!", or "Stop wasting time on Facebook!". The makers of Pavlok claim that this device helps unlock people's true potential, making them accountable for their behaviour and better able to change it when needed. This device relies on the idea that with the help of conditioning exercises similar to the famous experiment Ivan Pavlov performed on a dog at the beginning of the twentieth century, one can alter people's behaviour. Pavlok wearers testify that they were able to change their bad habits of overeating, nail baiting, hair pulling, and oversleeping because they started associating the feeling of being zapped with a prohibition on engaging in the bad habit.

Pavlok seems to be an ideal accessory in an era when external prohibitions linked to traditional authorities are on the decline, and when people are increasingly imposing prohibitions on themselves. The idea that people need to constantly work on themselves and engage in various forms of self-improvement is the basis of the majority of apps and wearable devices. The invention boom related to these devices has raised the question of whether one truly needs to control and measure so many things in one's life and what people gain with this multitude of apps.

Kerastase, a producer of hair care products, is, for example, planning a new hairbrush, designed together with the tech company Withings. This "smart" hairbrush is supposed to assess how people treat their hair. With the help of a built-in microphone, the brush will listen to how people style their hair and then try to determine how frizzy or dry their hair is and even whether they have split ends (Weatherford, 2017).[1]

Another example is Apple, who together with Nike created the Apple Watch Nike+, which comes in two sizes and features built-in GPS tracking, a perforated sports band for ventilation, Nike+ Run Club app integration, and exclusive Siri commands to start a run. On top of that, the watch is equipped with push notifications that are supposed to make us more prone to exercise. The Nike+ Run Club app entices wearers to run by offering daily motivations through smart run reminders. "Are we running today?", for example, appears on the watch at the time when the person usually goes for a run. The app also sends challenges from friends and even alerts runners about the weather outside. It is not just that training data, including pace, distance, and heart rate are available at a glance, one also shares run summaries with one's friends, which is supposed to promote friendly competition. The app even allows users to send fist bumps to each other right from the wrist as a form of encouragement.

Constant nagging, comparison with others, and even punishment are tactics that new technologies are adapting in their attempt to make people follow their life-improving plans. For some, these apps might be of help when they try to change their habits, but one should not forget that the whole ideology behind self-improvement, which is linked to ideas concerning choice and success, contributes to feelings of inadequacy, anxiety, and guilt. Paradoxically, these feelings are an important underside of the post-industrial ideology of choice, which stresses the power of individual rational choice and to a much lesser degree issues that are a part of social choices where the state and other power players are in charge (Salecl, 2011).

One way we deal with these unpleasant feelings is by taking the device not as a surrogate super-ego voice that is supposed to replace our internal super-ego injunction that makes us feel guilty, but rather as an object that somehow does the job instead of us. Austrian philosopher Robert Pfaller coined the term interpassivity in order to describe people's strategy when a device takes on the role of an intermediary that performs certain acts instead of the person (Pfaller, 2009). An example is a person who constantly records movies, but never watches them. By recording the films, it is as if the person is allowed to do other things while the recorder "watches" the movie for him or her.[2] Similarly, when a person makes a photocopy of a book that he or she never reads, it is as if the photocopier enjoys the book instead of the person.

One can take tracking devices as these kinds of objects that are doing the job for us. When I download a daily planner or a fitness app I can easily continue to not do the task I planned, since the app is a stand-in to which I somehow delegate the enjoyment of doing it for me. The very act of downloading is already an act of work, even a moment of sacrifice (if I had to pay money for it). After I have completed this task (and sometimes it does not go easy), I can for a short while play with it, but soon the very fact that I have downloaded it will be enough – I can go on doing whatever I am doing, while the app is supposed to do the work for me.

Let us take the example of a meditation app. I download it, maybe pay for it, do some meditations that it guides me to do, but in a few days I forget about it. Since I have it on my device, the app becomes a stand-in for my meditation practice. Invoking the term interpassivity, it can be said that the app is doing meditation for me, while I can go on doing my other things.

If I can easily forget the app that I have downloaded and if the very fact that I have downloaded it seems to be already enough for me to feel content with myself, the problem is that the app does not forget about me.

Ignorance and big data

While most of the discussions about wearable technologies focus on the question of whether they work or not in changing people's behaviour and how it is that people so easily ignore these devices, another form of ignorance – the one that pertains to the data that these devices collect about people – has been under far less scrutiny.

Amy Pittman recalls a time when she was trying to get pregnant and became enthusiastic about a period tracker. As she points out:

> Like many 20-somethings, I have an app for just about every important thing in my life. I have a health tracker that I ignore, a budget tracker that I ignore, an app to pay my bills that I try to ignore, and a period tracker that I'm obsessed with. Every week, I religiously tracked my mood on the period tracker along with my core temperature, the viscosity of various fluids, how often my husband and I were having sex, if sperm was present, etc. The app had more intimate knowledge of my reproductive behaviour than my husband or any doctor. On the day of my positive pregnancy test, I logged into my period tracker to share the good news. When I did, it suggested a pregnancy app, which I downloaded immediately. It was full of bright colours and interactive graphics.
>
> (Pittman, 2016)

Sadly, Pittman soon miscarried. At that moment, she deactivated her pregnancy-monitoring app. But logging off from the app did not prevent various marketing companies that target expecting women from continuing to send her info on pregnancy and baby products. The maker of her pregnancy app sold her info to marketing companies, however, when Pittman logged a miscarriage into the app and stopped using it, that information was not passed along. Pittman describes her shock when:

> Seven months after my miscarriage, mere weeks before my due date, I came home from work to find a package on my welcome mat. It was a box of baby formula bearing the note: "We may all do it differently, but the joy of parenthood is something we all share."
>
> (Pittman, 2016)

A whole new surveillance domain has opened up with the help of big data that allows commercial companies, as well as the state, to monitor people's daily lives. It is possible to ascertain that at the start of this massive collection of data people did not have an understanding of the market related to data collected about them. With various media addressing the problem of surveillance, however, it has become clear that it is not so much a lack of knowledge that is at work in the way people deal with their personal data, but rather a problem of ignorance.

The French psychoanalyst Jacques Lacan made a puzzling statement when he said that people do not have a passion for knowledge, but rather a passion for ignorance. Lacan observed that although his patients come to analysis with the desire to learn about what is guiding them in their unconscious, in the process of analysis they will go to great lengths to not come close to some traumatic knowledge (Lacan, 2007).

Sigmund Freud, in his own time, also established the importance of not knowing. He looked at the strategies of negation that people adopt when they deal with something traumatic. One of Freud's patients during his analysis described a dream and all of a sudden uttered: "The woman in my dream is not my mother" (Freud, 2001). What was surprising about this sentence was that Freud had not implied that the woman in the man's dream could be the patient's mother. The negation was coming from the patient, and with this negation the patient was naming something and saying that something is not true.

Freud's explanation is that we are dealing here with a repressed idea, which emerges into consciousness by way of being denied. Negation becomes a way of making cognisant what is repressed in such a way that it labels the repressed idea. Through negation we therefore in some way reveal traumatic truth, i.e. negation is the first sign of recognising that truth, but not yet accepting it, which is why we resort to denial.

For Freud, denial becomes both a testimony to the uncompleted task of recovering content from the repressed and some kind of a substitute for repression. It is, however, important to distinguish between denial and a lie. While a conscious lie would be an act to deceive, denial would rather be an act of impotence (Ver Eecke, 2006: 34).

When we deny something we inadvertently reveal what we wanted to hide. That is why denial also entails the opening up of a crack or fault, where a thought we were previously not conscious of suddenly emerges, which is why Freud, paradoxically, linked negation to the idea of freedom. He pointed out that when we have this crack, when through denial something emerges that is linked to traumatic truth, there is a possibility that the subject will start working through what had been repressed. It is, however, also possible that the subject will resort to new forms of repression.

How is big data related to denial? When we connect ourselves to all kind of apps and tracking devices, often the last thing on our mind is what will happen to the data such devices record. Often, it is as if one does not perceive data to be related to oneself or even that one does not think there is data collected and passed on in the first place. We thus simply ignore that information about ourselves is collected, and that this information might be sold and in various ways used for surveillance and marketing purposes.

Denial, however, becomes more complicated if we look at the content of the knowledge people often do not want to engage with. Studies in the field of denial have observed particular ways in which people deal with traumatic information in the domain of medicine (Dorpat, 1987). Shlomo Breznitz observed seven different kinds of denial among his patients (Breznitz, 1983). They often went from one type of denial to another; however, when a situation gets worse people often tend to regress to a more "primitive form of denial".

The first type of denial involves the negation of personal relevance. An example here is a study where a group of coronary patients witnessed a fatal cardiac arrest when they were in hospital. These patients, however, did not think that something like that could happen to them, too. The second is the denial of urgency. There is the example of people who in the past experienced a health emergency (a heart attack or cancer) and then delay calling for help when they experience a reoccurrence of the health problem. The third involves the denial of one's vulnerability. Here, the cases involve people who feel that because they changed their lifestyle (they exercise, eat well, etc.) they are somehow protected from having another health crisis (for example, a heart attack). Another form of denial of one's vulnerability involves people who completely give up their responsibility and perceive a heart attack as simply a matter of luck, fate, or other such uncontrollable factors. The fourth type of denial involves denial of the effects related to the traumatic experience that they went through. People who experience a heart attack might, for example, completely deny the anxiety related to this near death experience. The fifth type of denial involves people who experience certain affects and emotions in a life-threatening situation, however, they attribute them to other causes and not to the illness they are dealing with. Anxiety related to some rather insignificant issue can become a substitute for the life-threatening situation. The sixth form of denial involves the denial of information. A person might thus on a conscious level block any relevant information with regard to their illness and even disregard the advice they have been given by their doctors – i.e. coronary patients might stop exercising, not follow their prescribed diet, etc. On the unconscious level, however, they might very well have registered the information while they consciously deny it. The seventh form of denial pertains to severely depressed patients and to cases of psychosis where there might be indiscriminate denial of all information and the patient just seems to be in his or her own world, where the information regarding his or her health is simply not taken in. People might form delusions about their health that enable them to hold it together, however, their doctors' information about the illness is completely rejected.

With regard to big data and Internet security related to all kinds of information about us that is collected, we can observe a similar list of denials. Some people might have witnessed or read about cases of personal data being mishandled, but they do not think that something like that can happened to them.

Others might not be bothered at all that their data is passed to corporations or the state. Still others might be anxious that someone might be listening in to their phone conversations, but are not bothered that data about their life are recorded by a fitness tracker. A person might also have the delusion that there is a camera recording his or her daily activities, while he or she does not take in the problem that his or her data are being collected by mobile phone apps.

Nancy Tuana (Tuana, 2006) created a taxonomy of ignorance wherein she distinguishes four different ways we engage with the problem of not knowing:

1 knowing that we do not know, yet do not care to know;
2 not even knowing that we do not know;
3 not knowing because (privileged) others do not want us to know;
4 wilful ignorance.

If we apply this taxonomy to big data, we can observe all four ways of not knowing at work in the way people engage therewith. We might not know what the data collected about us is used for and not care about it. We might not know that we do not know what happens with the data. It is possible that companies that collect the data do not want us to know. And it is also possible that we resort to wilful ignorance, i.e. in this case, we know that data is collected, that it is sold, and that it can be abused, but we simply resort to not caring about it.

Another explanation of ignorance with regard to big data is that people are concerned that data is collected for potentially powerful uses that are not fully understood (Andrejevic, 2014: 1682). Here, ignorance does not so much pertain to the fact that data is collected, but to dealing with the question of how it is used. A person might thus be aware of the collection of his or her data, however, the word of corporations who traffic in this data, the mechanisms of data mining, and the working of algorithms is something so alien and opaque that he or she cannot envisage what such data can be used for or how it can be manipulated.

Jacques Lacan pointed out that ignorance is not misrecognition. When we misrecognise something, there has already been a level of knowledge that has first been recognised and then in the next step misrecognised (Lacan, 1988: 167–168). With regard to big data, misrecognition would be when we know that data is used in a manipulative way, but we misrecognise that as something beneficial. Ignorance, however, has to do with the fact that we close our eyes to knowledge that is too traumatic for us to bear. It might very well be that the opaque world in which data is used presents something so traumatic that we would rather close our eyes and do not want to come close to traumatic knowledge. Which is why we often so blindly consent to whatever Internet and app providers require us to do.

Informed consent

Whenever we download apps, sign up for free Internet in public spaces, register for loyalty cards, or put on wearable technology we are usually asked to tick a box that asks for our consent to the collection of data. Most often, we do this without reading the long document that in small print and in bureaucratic and legalistic language informs us of the rights of the service provider. We automatically click on the consent form and hope to start using the service without further interruptions. If we so easily ignore what we have given our consent to, one must question the purpose of engaging a customer in this game of consent.

The contemporary idea of informed consent originates in medicine (Murray, 1990). Its underlying presumption is that a person is a rational subject who can in an impartial way assess the information presented and then make a rational choice with regard to his or her well-being. The perception is also that a person who consents to a particular action has a clear understanding of the consequences and implications of such action.

The idea of informed consent historically emerged as a result of various forms of abuse that happened in the domain of medicine. The most important were the medical experiments performed in Nazi Germany on prisoners in concentration camps. Cases where people were either deceived or coerced to take part in medical research or when people were not informed of the possible outcomes of certain medical procedures contributed to the demand that consent become an important part of the interaction in the domain of medical practice and research. Struggles against paternalism in medicine, as well as appeals for respect for the autonomy of the patient have also contributed to the promulgation of the idea of informed consent (Manson, 2007).

Medical ethicists discuss many dilemmas related to informed consent – from the question of what it means to be properly informed, to the capacity to make decisions, legal aspects of consent, to cases of the exclusion of consent (with regard to children, mentally disabled patients, etc.). Rarely, however, do discussions touch on the conscious and unconscious mechanisms that guide people in their decision-making and also in their refusal to make such decisions. What is equally neglected in discussions on informed consent is the embracement of ignorance on the side of the patient as well as on the side of the authority in charge of drafting the consent form (e.g. a medical institution).

For an informed consent to become a viable legally enforceable contractual document, it needs to encompass a certain perception of the subject as a rational person capable of making decisions that contribute to his or her well-being. Both the illusion of rationality as well as the illusion of the utilitarian tendencies of people underpin the idea of informed consent. Dilemmas related to the unconscious mechanisms that guide people, as well as the fact that people often do not follow utilitarian ideas about maximising pleasure and minimising pain had to be refuted.

This illusion of rationality is a necessary prerequisite for the establishment of the contractual relationship between the patient and the doctor (as well as the medical institution). In our highly litigious times, informed consent, however, has opened the doors to new forms of ignorance on the side of the patient. When we undertake the most insignificant medical procedure, we need to sign a document where we agree to all kinds of possibly damaging outcomes of these procedures. One usually quickly glances over the text and signs the form without actually fully rationally digesting the information presented. Here we embrace various forms of denial, which are not so different from the above-mentioned denials.

We might engage in wishful thinking that all the disasters that can happen during the procedure will not happen to us. We might also deny the effects that the information on such disasters provokes in us. Or we might deny that there is a rational logic presented in the document as such. Since we know that informed consent forms are cut and paste documents that are used for various situations, we might perceive them as merely legal gibberish that acts as a protective shield for the medical establishment.

Without this ignorance, it is quite possible to envisage that a person who took seriously the warning as to what might go wrong as presented in the informed consent document would not choose to undergo the procedure or might become extremely anxious or even paranoid.

A similar situation is at work in our dealings with the Internet. If we were to read all the various informed consent documents that we blindly agree to, it is quite possible that we would not install the majority of apps on our phones, put on wearable technology, or connect to open Internet servers.

The problem with informed consent is that it primarily protects the provider of a service, while for the consumer it more and more presents a case of a forced choice. We are offered a choice to either consent to giving away our data or not. However, if we say no, we lose the very possibility to enjoy the device that collects the data. Similarly, if we do not consent to allow ourselves to be monitored by Internet providers we are denied connection to the Internet in the first place. In cases of forced choice, one is in principle offered a choice, however, this choice involves only one option. In a way, choice is offered and denied at the same time.

An example of forced choice existed in socialist Yugoslavia when young men were obliged to serve in the army. When young men became conscripts, they had to go through a ritual where they took an oath saying that they freely chose to become a member of the Yugoslav army. However, one man took this choice seriously and said that since becoming a member of the army was a matter of choice, he chose not to join it. When this happened, he was immediately sent to prison. The choice in question was offered and denied at the same time.

Lacan explained the idea of forced choice by envisaging a situation wherein a man is confronted by a robber who demands: "Your money or your life!" This demand puts the man in a position of forced choice. If he chooses his money, he will lose his life and thus will not be able to enjoy the wealth that he saved. The only choice that is left to him is his life, which, however, will be less enjoyable since he will lack money (Feldstein, Fink, & Jaanus, 1995: 47).

Similarly, when we are asked to consent to the use of devices that track our data, we are offered a choice: enjoy our app, but give us consent to do with your data what we will or you can have your life without the app. The choice is thus between life without data or digital death.

Machines cannot be wrong

While we blindly consent to giving our data away, we often also blindly place our trust in the machines that handle such data. Belief in the power of computers is such that we often do not even envisage that serious mistakes can be made in the way they work.

A few years ago, I presided over a panel that evaluated the output of research groups. I was not linked to these research groups and the evaluators were from abroad. This setting was supposed to allow for an objective account of the researcher's work, which, of course, had serious implications for their future funding. My job was fairly simple. On top of facilitating the evaluators' reports, I had to put their marks into an Excel spreadsheet, which in the end would automatically calculate the results, providing me with a list of winners and losers. I meticulously recorded the marks into Excel so that potential errors might not affect the results. In the end, I got the results and the evaluation was done. A few hours later, I looked at the form again and had the feeling that something was amiss. Groups that consistently got good marks from the evaluators were not as high on the list of results as I had expected. I rechecked whether I had put all the marks into the form correctly and it all looked fine. I clicked the calculation button again and got the same results as before. Frustrated, I decided to do the calculation by hand. To my surprise, the results turned out different. I denied the possibility that the computer might be wrong and decided to do the calculations one more time. Finally, I had to acknowledge that the spreadsheet had not been formulated properly. When I contacted the agency that had set up the Excel spreadsheet at first no one believed me that the machine had produced the wrong calculation. Finally, the IT personnel confirmed that there was an error in the algorithm which, as a result of my complaint, they were able to solve. Until I had that experience, I was a very trusting user of similar forms. Subsequently, I started wondering how many similar calculation mistakes are at work in our computer-dependent work and why we do not pay more attention to them.

In the world of big data, we must not only deal with potential computer failures, but also a high level of opacity related to how this data is collected, how it is interpreted, who has access to it, and how it can be manipulated. We also deal with sample bias, as well as an increased desire to see in data what we want to see in the first place. In addition, the way companies use algorithms to comb through data is usually secret.

It is thus not surprising that big data is opening new avenues of blindness. Paradoxically, when we collect a great amount of data, suddenly people start seeing patterns in random data. Researchers of big data thus point out that we are experiencing apophenia: seeing patterns where none actually exist, simply because enormous quantities of data can offer connections that radiate in all directions (Dugain & Labbe, 2016).

One of the ways we often deal with blind spots is by trying to visualise them. "Gaps", cracks in knowledge, are in a particular way linked to the fantasies we create around them. Art provides one way to look at these gaps. Contemporary art has been fascinated with the new developments in science. We can thus find numerous artists who use brain images, genetic code, and knowledge from the fields of astrophysics and physics in general in their art. Not surprisingly, big data has also found its place in the domain of art. The Norwegian artist Toril Johannessen, for example, in her art project "Words and Years" uses big data to create pictures that try to alert viewers to important themes in today's world. Combing through data in scientific journals, she created a picture that shows when and with what frequency the word crisis is used with regard to nature or society, how often the word miracle is used with regard to nature and society, and how many articles in the field of genetics deal with the words greed and desire.[3]

Even before the rapid development of statistics, artists collected data and used them in art works. In the 1980s, the Russian artists Vitaliy Komar and Alex Melamid, for example, conducted surveys in different countries asking people what a beautiful painting looks like and what an ugly one looks like. Following the results of the survey, they then produced an ideally beautiful painting and an ugly one. Quite universally, the result of the surveys was that people perceived as beautiful paintings those that showed scenes of nature with mountains, with a sunny sky, and an animal in the setting, while they perceived as ugly paintings that consisted of abstract triangles in dark, unappealing colours. Both the most beautiful and the most ugly painting used the mean results of the artists' surveys (Dissanayake, 1998). Observing the beautiful and ugly painting that they then painted incited in the viewer an uncanny feeling – trying to comply with people's idea of what beautiful or ugly painting looks like took away the edge of surprise that often accompanies good art work. By taking seriously what people perceived as beautiful and what as ugly art, the artists tried to put into words and realise in an image what usually cannot be grasped in a rational way. What makes an art work great usually escapes words, which is why it is not easy to rationally describe what makes one art work beautiful and another ugly.

In order to depict the nature of what cannot easily be put into words and what escapes people's rational perceptions of themselves as well as the world around them, Jacques Lacan used the term "the real". This term does not pertain to what we usually understand as reality, but rather to what escapes the perception of reality that we form with the help of language as well as images.

In today's times we have various attempts to come close to this real with the help of science and new technology. Genetics and neuroscience, for example, give us the perception that decoding the genome and perfecting brain scans might help us comprehend what makes us human. Big data, in its own way, tries to closely approach the secret of subjectivity. These attempts, open up the space for new fantasies to be formed around the ungraspable in subjectivity. Dominique Cardon points out that we need to ask what algorithms dream about and how they operate on human desires. Although we are often under the impression that with the help of algorithms we can escape the "tyranny of the centre" and enable the diversification of society, which as a result will hopefully be less hierarchical, in reality the opposite is true and algorithms allow for the perpetuation of inequalities. And we should not forget that the devices that provide data are also becoming objects with the help of which new forms of hacking attacks can easily become carried out.

One of the most surprising cyber attacks happened in the US in 2016 when a large number of security cameras and other domestic devices were infected with a fairly simple program that guessed their factory-set passwords – often "admin" or "12345" or even "password". Once these devices were infected, they were turned into an army of simple robots, which at a coordinated time were then instructed to bombard a small company in Manchester, NH, with messages. This attack overloaded the circuits of the company, which functioned as one of the Internet's giant switchboards, so as a result of its failure numerous US companies such as Twitter, Reddit, Airbnb, and even the *New York Times* lost their Internet connection or it slowed to a crawl (Sanger & Perlroth, 2016).

While many experienced this Internet attack as a fleeting inconvenience, it portends much more. In the era of the Internet of Things, the problem is not only that such hacks happen in interconnected refrigerators and security cameras, but that they are also happening to a growing number of medical instruments and recreational devices – such as heartbeat-monitoring watches – that report medically relevant information. Cybersecurity for these devices is increasingly becoming a big problem, since state regulations that pertain to financial data often do not pertain to health care records (Haun & Topol, 2017). Taking into account the fact that these devices are portable, one encounters problems related to their security that go beyond state jurisdictions. Which is why some cybersecurity experts are calling for industry-wide cooperation in the adoption of security standards before some major hack occurs, while others are trying to teach people how to protect themselves from having their devices hacked and their private data appropriated by new types of cyber criminals.

Conclusion

We often glorify the pursuit of knowledge; however, the desire to not know is equally important for our survival. Closing our eyes, not seeing something that is traumatic, or not remembering what has been painful and hard to deal with are strategies that people have embraced with a passion equal to that with which they have embraced the pursuit of new knowledge.

In order to understand the nature of such ignorance, we can say that it in some way allows a person to not come close to what is traumatic. In our private lives, repression helps us push away what is for us consciously hard to comprehend. But with ignorance, it is as if we have all the information but it does not pertain to us. An individual, for example, can have information about a threat, but behave as if it does not concern him. This kind of ignorance paradoxically contributes to a feeling of omnipotence; we perceive ourselves as being more powerful than we actually are. Such feelings of omnipotence can contribute to belief in the idea of technological development and progress that does not allow for seeing the negative consequences thereof.

If we compare ignorance of the use of big data with the denial that can be observed with regard to climate change, we can see a similarity at work in the way these two forms of closing our eyes deal with the idea of progress. People in the developed world are afraid to admit that the belief in development that underlies modern capitalism is in fact something that cannot last forever. People are also afraid to face the prospect that climate change might actually lead to a decline in economic growth, and that any government intervention in the market through various mechanisms of controlling carbon dioxide emissions and introducing penalties for corporations might also imply loss of the idea of freedom, which for many people is related to the idea of the free market. Even those who are aware of the warnings that scientists are issuing as to climate change often have various strategies enabling them to believe that these warnings do not affect them per se. People often deny both that climate change means that they themselves need to do something and that society needs to change its course as regards what it perceives as development.

This denial is often related to the fact that a lot of people are afraid of change and that they are anxious about what potential changes might mean as regards their future. People might also be afraid that the future will not involve the idea of progress, which they hope will continue. Clive Hamilton warns that the climate change bill will be received by the next generation, since it will entail a bill issued for the incredibly rapid development of the past, which is based mainly on energy that we obtain from fossil fuels (Hamilton, 2015). Prosperity is important for the current generation, something that allows this generation to live longer, healthier lives in the developed world. The problem of this generation, however, is that it has not paid the full price of this progress. The rest of the price will be charged to future generations.

With regard to big data, we also have an over-optimistic idea of how this data contributes to progress. Here, too, the price for this belief will be paid by future generations. On top of problems involving the mismanagement of data, bombardment by consumerism, and new forms of surveillance, future generations will need to deal with the fact that it never consented to its data being collected from the moment of conception. Which is why researchers dealing with the problems of big data warn that our ideas of privacy and informed consent do not encompass the fact that data on children are nowadays being collected on a massive scale without them being able to control or comprehend the impact this will have on their future lives (Lupton & Williamson, 2017).

Optimistic big data researchers like to point out that big data need not be regarded simply from a negative perspective, i.e. people can be empowered to use data to their advantage and that data that is available through open access can significantly contribute to scientific research and social change.

Proponents of big data thus like to point out that an individual should have access to his or her data so that he or she can make full use of it. The idea is that a person's tracking devices and computer know more about his or her habits than he or she consciously does. Knowing about the data that is collected about oneself will help one better navigate life. Mark Andrejevic warns against such enthusiasm by pointing out that there exists a great discrepancy in power between those who collect big data and those who are the objects of such collection:

> Even if users had access to their own data, they would not have the pattern recognition or predictive capabilities of those who can mine aggregated databases. Moreover, even if individuals were provided with everyone else's data (a purely hypothetical conditional), they would lack the storage capacity and processing power to make sense of the data and put it to use.
>
> (Andrejevic, 2014: 1674)

One can add to this the observation that psychoanalysis, already at the time of Freud, observed that, sadly, people might rationally state that they are concerned about their well-being while unconsciously they do everything that goes against this idea. They thus often do not follow their rationally proclaimed goals, but rather continue on the path of pain, guilt, and even self-punishment.

Notes

1 The explanation is that the brush works in such a way that "three-axis load cells measure the pressure you exert on your hair and scalp as you brush, and sensors count the number and speed of brush strokes, and gauge if hair is being brushed wet or dry" (Weatherford, 2017).

2 Pfaller expanded his theory to works of art. In exhibitions of contemporary art it often happens that the visitor does not have an idea what the works of art he or she is observing are all about. When walking around the exhibition, the person, however, can have the impression that the curator somehow viewed the exhibition for them (Pfaller, 2009).

3 www.toriljohannessen.no/Words_and_Years_page_1.html.

Bibliography

Andrejevic, M. (2014). Big data, big questions: The big data divide. *International Journal of Communication*, *8*(0), 1673–1689

Baumeister, R. F., Bratslavsky, E., Muraven, M., & Tice, D. M. (1998). Ego depletion: Is the active self a limited resource? *Journal of Personality and Social Psychology*, *74*(5), 1252–1265.

Breznitz, S. (1983). *The denial of stress*. New York: International Universities Press.

Burkeman, O. (2017, 7 January). How to keep your resolutions (clue: it's not all about willpower). *Guardian*. Retrieved 8 January 2017 from www.theguardian.com/lifeandstyle/2017/jan/07/how-to-keep-your-resolutions-not-all-about-willpower.

Cederström, C., & Spicer, A. (2015). *The wellness syndrome*. Cambridge: Polity.

Clear, J. (2016, 11 January). *The akrasia effect: Why we don't follow through on things*. Retrieved 8 January 2017, from http://jamesclear.com/akrasia.

Dissanayake, E. (1998). Komar and Melamid discover pleistocene taste. *Philosophy and Literature*, *22*(2), 486–496.

Dorpat, T. (1987). A new look at denial and defense. *Annual of Psychoanalysis*, *15*, 23–47.

Dugain, M., & Labbe, C. (2016). *L'homme nu. La dictature invisible du numérique*. Paris: Plon.

Elster, J. (2016). *Sour grapes: Studies in the subversion of rationality*. New York: Cambridge University Press.

Feldstein, R., Fink, B., & Jaanus, M. (1995). *Reading seminar XI: Lacan's four fundamental concepts of psychoanalysis: The Paris seminars in English*. New York: SUNY Press.

Finkelstein, E. A., Haaland, B. A., Bilger, M., Sahasranaman, A., Sloan, R. A., Nang, E. E. K., & Evenson, K. R. (2016). Effectiveness of activity trackers with and without incentives to increase physical activity (TRIPPA): A randomised controlled trial. *The Lancet Diabetes and Endocrinology*, *4*(12), 983–995.

Freud, S. (2001). *The complete psychological works of Sigmund Freud, Vol. 19: "The Ego and the Id" and Other Works*. London: Vintage Classics.

Hamilton, C. (2015). *Requiem for a species*. London: Routledge.

Haun, K., & Topol, E. J. (2017, 2 January). The Health Data Conundrum. *New York Times*. Retrieved 9 January 2017 from www.nytimes.com/2017/01/02/opinion/the-health-data-conundrum.html.

Lacan, J. (2007). *Ecrits: The first complete edition in English*. (B. Fink, Trans.). New York: W. W. Norton and Company.

Lupton, D., & Williamson, B. (2017). The datafied child: The dataveillance of children and implications for their rights. *New Media and Society*, 1–15. Retrieved 10 January 2017, from http://journals.sagepub.com/doi/pdf/10.1177/1461444816686328.

Manson, N. C. (2007). *Consent and informed consent*. Retrieved 8 January 2017, from https://repository.library.georgetown.edu/handle/10822/968229.

McGonigal, K. (2013). *The willpower instinct: How self-control works, why it matters, and what you can do to get more of it*. New York: Avery.

Murray, P. M. (1990). The history of informed consent. *The Iowa Orthopaedic Journal*, *10*, 104–109.

Pfaller, R. (2009). *Ästhetik der Interpassivität*. Hamburg: Philo Fine Arts.

Pittman, A. (2016, 2 September). The Internet thinks I'm still pregnant. *New York Times*. Retrieved 8 January 2017, from www.nytimes.com/2016/09/04/fashion/modern-love-pregnancy-miscarriage-app-technology.html.

Salecl, R. (2011). *The tyranny of choice*. London: Profile Books.

Sandel, M. J. (2013). *What money can't buy: The moral limits of markets*. New York: Farrar, Straus and Giroux.

Sanger, D. E., and Perlroth, N. (2016, 22 October). A new era of Internet attacks powered by everyday devices. *New York Times*. Retrieved 11 January 2017, from www.nytimes.com/2016/10/23/us/politics/a-new-era-of-internet-attacks-powered-by-everyday-devices.html.

Tuana, N. (2006). The speculum of ignorance: The women's health movement and epistemologies of ignorance. *Hypatia*, *21*(3), 1–19.

Ver Eecke, W. (2006). *Denial, negation and the forces of the negative: Freud, Hegel, Lacan, Spitz, and Sophocles*. New York: SUNY Press.

Weatherford, A. (2017). *This new hairbrush is totally judging you*. Retrieved 5 January 2017, from http://nymag.com/thecut/2017/01/kerastase-and-withings-created-a-smart-hairbrush.html.

4 Machines, humans and the question of control

Zoran Kanduč

General, your tank is a power vehicle.
It smashes down forests and crushes a hundred men.
But it has one defect:
It needs a driver.

(Brecht 1938: 289)

Introductory remarks: the digital age and its discontents

Although the term "big data" (written with or without capitals) may seem quite vague and elusive, it is, nowadays, widely and pompously used, particularly in post-modern scientific discourse (but not – or at least not yet – in ordinary language). Moreover, it often functions as some sort of master signifier designating a veritable paradigm shift in numerous areas of collective activities and individual lives. To put it more precisely, that means that "we" have entered into (or that we have been invaded by) a data-driven, digitalised economy, society, state and, consequently, control. Yet, what is really a new phenomenon is not the brute existence of enormous or even infinite quantities of all kinds of data or information. Namely, the incessant flux of undifferentiated impressions, perceptions, and other stimuli generated by "our" inner and outer (subjective and objective) worlds is a constant and inevitable feature of the human condition and the main cause of the necessary and regular use of various filters that select, organise, neutralise, dilute, or rarefy countless data (which otherwise would threaten the individual mind with suffocation or paralysis).[1] So, what is the assumed historical novelty that the term "big data" attempts to capture? Well, the answer is not a big secret. What is at issue is the appearance of ever evolving or progressing technology that is able to register, collect, store, mine, analyse, and otherwise manipulate or use huge amounts of data of various of formats at astonishing speeds.

Clearly, what we have to deal with is the relationship between human beings and technology, in other words: the very old problem of the use and abuse of (mainly natural) science and technology. To put it more directly, is the ever-expanding occurrence of "big data" technology (and the accompanying analytical approaches, skills, and knowledge) a bad or a good thing? Is it something that we should welcome with admiration, enthusiasm, or hope, e.g. due to its potential to solve persistent social problems? Or, alternatively, should it provoke doubts, fear, or even passive or active resistance? But can we really fight against scientific and technological progress, for it is seemingly unstoppable,[2] beyond political, social, legal, and moral control? In any case, what is or could be wrong with using advanced "big data" technology? Apparently, due to the combination of the immense power of modern computing with the plentiful information generated by the digitalised society one can make better, more intelligent, or at least well-informed decisions or choices. One can predict the (uncertain) future (or rather envisage a number of possible futures) with greater accuracy. One can spot hidden or potential business trends. One can detect "hot spots of crime" (i.e. extreme risk areas that should be flooded with targeted patrols). One can uncover terrorist plots and prevent such pernicious attacks. One can identify (and punish) those who cheat the social state, i.e. parasites living and even enjoying their lives to the detriment of the socially responsible, honest, and hardworking ("active") citizens. One can increase productivity or improve efficiency (of blue, pink, and white-collar employees). One can boost one's competitive market position. One can produce all sorts of "actionable" data. One can replace "all too human" judgements, assessments, rankings, ratings, or scorings of other people in various social roles and situations with much more objective and fair algorithms or automated ("artificially intelligent") systems that function without subjective (implicit or explicit) biases, deficiencies, prejudices, emotions, mistakes, or inconsistencies. One can predict the outbreak of a dangerous disease. One can find and buy a desirable commercial product or service at a lower price (e.g. private car transfer, tourist room, or airplane ticket). One can earn some money by participating in the so- (but mostly erroneously) called "co-operative" economy (e.g. as an Uber taxi driver or an Airbnb "hotelier"). One can try to get a sex partner or even a suitable person for a "serious", long-lasting love relationship. And so on, and so on.

So, should we conclude that "our" world (or at least its "free" and "democratic" part, belonging to the most developed, powerful, and rich countries) increasingly resembles a technological paradise, the realisation of the most audacious dreams of "fallen" humanity (or even the historical return of a mythical "golden age")? Well, what we can observe is that ordinary (more or less "little") people – not to mention their economic, financial, and political masters – evidently love and adore all kinds of commercial gadgets and machines, modern and post-modern, "stupid" and "smart" ones. Just think of their emotionally warm (or even amorous) attitudes towards cars, television sets, radios, mobile phones, personal computers, etc. Evidently, post-modern worshippers of technology (or believers in technical "progress") are thoughtful enough to imagine some useful or at least joyful way of using the newest gadgets (and of wasting their time, energy, and money by following and loyally obeying the dictatorship of the capitalist supply of market "goods"). What is more, there has been no massive protest against the use of "big data" by private employers (capitalist bosses) and state officials. Workers accept even the most humiliating forms of surveillance, spying, or control. So, it seems that there is nothing wrong with "panoptic" electronic monitoring and wearable devices tracking and analysing (by means of software algorithms) their tone of voice, body language, movements, geographic position, habits, (un)healthy lifestyle, quality of sleep, mental states (such as the level of "objective", measurable stress, fatigue, or happiness), interaction and communication patterns, drug use, blood pressure, hormones, heartbeat, etc. All these degrading aspects of monitoring (e.g. socio-, psycho-, and biometric tracking) pass as more or less normal phenomenon, as the taken-for-granted continuation of the "all too old" game of controlling hired workers, yet nowadays in new, extremely sophisticated forms. As something that employers are entitled to practise, for they have a right to make good and full (or rather profitable) use of the relevant bought commodity (labour power) – in the same way as the consumer has an indisputable right to squeeze all the juice from an orange purchased in a shop. On the other hand, also Snowden's revelations of the widespread (pervasive and consequential) phone-tapping program of the NSA were not met with serious public outrage. The same holds true for the similar actions of many other whistle-blowers or (old and new) media as well. As though uncovering the legally or morally problematic activities ("scandals") of rich individuals, corporations, or state officials were something that nowadays attracts the interest (and reactions) of just a thin minority of the professional or laic public. Therefore, it comes as no surprise that also critiques of "big data" technologies are not particularly influential. Yet, this does not mean that they are pointless. Quite the contrary.

Many critics point to the threats connected with the widespread use of algorithms (abstract instructions designed to produce specific outputs). They worry because these mysterious, opaque, invisible "judges" are deciding the destinies of an increasing number of human beings, e.g. job candidates,[3] workers, perpetrators of criminal acts, prisoners, loan or health insurance applicants, recipients of some sort of social help (or welfare recipients), consumers, tenants, entrepreneurs, investors, researchers,[4] etc. The problem is not solely that these digital "decision makers" lack empathy or that they can go awry (and in doing so, destroy one's reputation, savings, and investments, reduce study opportunities, diminish employment options, or falsely identify an innocent person as a would-be or actual criminal or terrorist). These computerised recipes for data processing, decision-making, or finding solutions are not so objective, neutral, or scientifically verified mechanisms as they perhaps appear at first glance. Namely, algorithms do not fall from the blue sky. They are not a gift from God. Clearly, humans make and remake them. Although the persons in question are highly qualified technical experts, e.g. computer programmers, data analysts, mathematicians, and statisticians, they are not immune to subjective perspectives, desires, prejudices, biases, misunderstandings, or stupidity pure and simple (as a universal and incorrigible component of our "all too human" nature).[5] Moreover, we should keep in mind that it is still humans who decide that algorithms will be applied in a particular field of action, e.g. due to their superior efficiency, reliability, or cost-effectiveness.[6] On the other hand, some commentators protest against the growing trade in personal data (digital "profiles" or "traces") that technological giants (e.g. Google and Facebook) sell to various clients (such as advertising, marketing, and insurance companies). Others are more concerned with the protection of "personal privacy" and "data privacy" (associated with an individual's social or psychological profile or fragmented versions of him- or herself made by automated systems and available to state agencies or private companies), for they assess that both are in danger in the contemporary "surveillance society".[7] And so on, and so on.

Numerous critical analyses focus on the use and abuse of "big data" in the extremely broad and differentiated field of surveillance. All too often, the crucial threat (e.g. to individual freedom, civil liberties, human rights, privacy, intimacy, or even democracy) is located in the post-modern ("automated") state, presumably increasingly resembling the worst Kafkaesque nightmares or almighty, all-knowing, all-hearing, and all-seeing Big Brother, as depicted in Orwell's notorious dystopia. Undoubtedly, computerised surveillance (based on data gathering and processing) has impressive repressive potentials. Yet, the efficiency of state ("top-down") control (e.g. in terms of transforming its citizen into totally transparent, glass-like items)[8] should not be overestimated or overstated, not only due to thc "all too human" bureaucracy, but also because contemporary "control problems" have become much more complex (difficult and, in many cases, even impossible to handle) than in the modern past.[9] Just think of the immensely heterogeneous mass of risky population embracing all kinds of "ordinary" criminals, "illegal" economic and other international migrants, asylum seekers, internal ("home-bred") and "imported" terrorists, members of organised crime, "black bloc" anarchists, religious (unassimilable) fundamentalists, annoying beggars, drug addicts, work-shy (insufficiently "active") plunderers of the dismantled social state, wanton or drunken car drivers, *indignados*, etc. In the post-war period (i.e. the Cold War), the image of inner and outer "public enemies" – both criminal and political threats (e.g. pro-socialist or pro-capitalist dissidents) – was much clearer.[10] Although the post-modern state is becoming more authoritarian, more and more citizens doubt its ability to ensure even elementary safety, so they feel forced to protect or fortify themselves by their own means (or by buying the goods and services of the blossoming private security industry). Moreover, what seems much more fearsome is not the controlling power of repressive state apparatuses, but the notorious weakening or debility of the postmodern state in relation to the "irresistible" dictatorship of the financial markets, multi- or transnational corporations ("global players"), and institutions of the "supra-national state of capital" (Gorz 1999: 15), e.g. the WTO, the IMF, the World Bank, and the OECD.[11] In the context of intensifying competition in global markets, the extreme mobility of capital (particularly speculative and financial capital), and the increasing orientation to short-term profitability, the state has allegedly no other alternative but to strictly obey the pitiless dictates of capital (as the sole remaining possessor of sovereignty). It has to be competitive itself and follow capital's demands for subsidies, tax breaks, infrastructure, workforce flexibility (disciplined, competent, and cheap workers), privatisation, deregulation (the elimination of all legal and administrative "rigidities" or barriers), lower public expenditure and reduced social costs (associated with welfare programmes and institutions).[12] In short, the old sovereign state, firmly confined to its territory, is obliged to adopt and subordinate its laws (and politics) to the "free" operation of the quasi-natural, anonymous laws of the market as the key, omnipresent, almighty, controlling mechanism.

Consequently, fears related to the oncoming quasi "totalitarian" state, equipped with unprecedented, historically unparalleled monitoring power, are not well founded. Instead, we should study the problem of post-modern technology-based (and, of course, other types of) control primarily from the perspective of the capitalist economy (as a world system). In other words, what matters most is the role of machines (products of successive industrial revolutions) in the perpetuation of capitalist class rule, and the exploitation, oppression, humiliation, and extortion of the structurally subordinated sellers of their labour power.

Capitalism, machines, and control

Let us begin with a perplexing paradox, formulated as a question. How is it possible that increasing gains in productivity (due to the astonishing progress of modern science and technique) have not diminished the painful burdens of work, i.e. of economic activity, which is – and will always be – absolutely necessary to satisfy the recurring individual and collective human needs and desires? Unfortunately, what happens is almost just the opposite. Generally, people still work too much. They routinely ruin their bodies and souls by working too long hours, under acute or even chronic stress, in degrading, unhealthy, or "dehumanising" conditions, under torturing pressure, in a great hurry or with the suspicious help of legal or illegal drugs. More often than not, they work in fear that they will lose their "precious" jobs (temporary employment or clients), even one that they do not like or actually hate, because it is poorly paid, fatiguing (physically, mentally, or both), annoying, lacking in dignity, respect, or recognition, stupefying, humiliating, boring, frustrating, or simply meaningless.[13] In the light of the prevailing neoliberal mentality, work whatsoever is superior to non-work ("passivity"), even if the acquisitive activity in question is undoubtedly harmful or destructive (for humans, society, culture, or nature).[14] Even poorly "rewarded" toil or a gratuitous ("voluntary") job is supposedly better than no work, for it helps maintain an individual's correct attitudes and work habits. What is still a "mortal sin" (an economic, moral, or political problem) is laziness ("non-activity"). That is the principal "vice" to be blamed, punished, or otherwise prevented.[15] As a matter of fact, we are witnessing a bizarre historical irony. Namely, in the dramatic ("revolutionary") period of "the great refusal" (in the 1960s and 1970s), all key pillars of modern (industrial) capitalism were under attack, including (or even principally) heteronomous work.[16] Nowadays, it is precisely work that people demand and fight energetically for.[17] This is not at all surprising. There is not enough paid, profitable, or "productive" work ("jobs") for all humans who desperately need money (and therefore an employer, or possibly self-employment) in order to buy the various commodities necessary for living. Evidently, increasing amounts of "live" work (understood as a specific "social construction" peculiar to the modern, industrial capitalist system) is "saved", i.e. irreversibly eliminated or abolished in all sectors of the economy and at almost all levels of production. Under the "irresistible" impact of information technology, automatisation, digitalisation, use of robots, or algorithmic regulation, a growing number of unqualified and also (highly) qualified workers become unemployed (or under-employed), "downsized", redundant, or part of the swollen superfluous population that can earn money solely in the "informal" (or criminal) economy. What is worse, it is not just "work" that is disappearing, but also the "normal" job (regulated strictly by labour law and functioning as the foundation of other important rights) or the "classic" wage relationship (as "we knew it"), for the normative status of post-modern acquisitive activities (and sellers of labour power) is highly differentiated. Just think of project, contract, temporary, sessional, "illegal", emigrant (with papers), agency, informal, and self-employed (freelance) workers.

To repeat the paradox once again, more and more wealth is (and can be) produced with less and less work. Yet, the enormous productivity (as a result of the incessant and even vertiginously accelerated progress of labour-saving technology) functions not as a blessing, but as a curse on the increasing number of humans who have to sell their labour power in order to survive or live more or less normal or "decent" lives (according to the normative expectations of their reference groups). Globalised, neoliberal, technologically advanced capitalism has caused an unprecedented gap between the super-rich ("winners") and the rest of humanity (the relatively and absolutely poor),[18] deeply rooted social polarisations, massive unemployment, generalised workers' insecurity, uncertainty, and fear (not to mention ecological threats and the already existing damage and deterioration of the natural environment). On the other hand, the disappearance – or "end" (Rifkin 2007: 15–27) – of work, in the context that is still firmly structured and conceived of as "*Arbeitsgesellschaft*" (Arendt 2016: 398–404), has enviable advantages for capitalists in particular and for employers generally. Namely, capital is increasingly freeing itself from all normative and factual limits (coming near to the ideal of perfect liquidity). Beck (2003: 15–20) points out that entrepreneurs (especially transnational corporations) have discovered "the philosopher's stone" for accumulating wealth and, consequently social, political, and legal power: capitalism without work and without taxes. Employers can select workers from the vast ocean of eager candidates competing for their favours. Economic masters, as the indisputable winners in the class war over structurally subordinated labour, can hire labour power "just in time", only when they need it (and dismiss it after profitable use). In other words, they can pass the burden (or "violence") of intensified competition on to economically dependent "partners", i.e. to workers and subcontracting companies. "Happily" employed staff have no other option than to increase or maintain productivity at the demanded (global) level or to accept longer working hours, lower payment, and fewer traditional (nowadays apparently obsolete) rights.[19] Therefore, it is not an exaggeration to describe the hyper-accelerated postmodern world as a paradise for the ruling class and as a hell for the ruled, subjugated, insecure (precarious) workers, servants, wage "slaves", vassals, and especially those deprived of the "privilege" of actively participating in capitalist exploitation.

How to explain this socioeconomic and political "miracle"? Is there anyone or anything in particular that we can blame? Well, to begin with, we cannot accuse labour-saving technology as such. On the contrary, machines that save, alleviate, or simplify irreducible human work contain obvious liberating potentials. Their use, accompanied by a rational, co-operative, and humanised organisation of work and social (democratic) control of production, could dramatically reduce "the kingdom of necessity" in favour of "the kingdom of freedom." In theory at least, although in capitalist practice this does not happen. Yet, we cannot criticise capitalism for that reason. Namely, its main purpose is not to promote social welfare, individual freedom, human rights, and democracy, nor to provide enough wage work for all. The capitalist economy is inherently irrational and destructive (particularly when its development is not limited "from the outside", i.e. by the opposition or resistance of workers and by the power of the state). Its basic – or even sole – purpose is the valorisation of capital, the production of surplus value, infinite and limitless accumulation (the transformation of extracted surplus value, realised in monetary form, into capital). Capitalists use human workers (as formally free and equal legal subjects, i.e. owners of their labour power) and a natural resource solely as means in their "never ending" race for profits. However, they are forced to do so, if they want to stay (i.e. avoid economic ruin) and prosper in the market "game". Also, capitalists are compelled – by the inexorable "law of competition" – to constantly increase the productivity of labour (by co-operation, the division of work, and the use of machines),[20] preferably to gain some extra profit, at least temporarily (before the technological or organisational innovation becomes an obligatory social standard or average condition of production).[21] In that regard, capitalism functions as a system of mutual pressures and threats. Each capitalist is under constraints and he or she, at the same time, constrains all other competitors. Consequently, everyone has to follow the "blind force of things" (Heinrich 2013: 110).[22] That is why capitalism looks like a gigantic machine, running without some ruling, directing machinist (usually written with a capital letter) or collective mastermind. There is no one whom we could consider responsible for the enormous human, social, and ecological damage (or destruction) caused routinely ("naturally") and legally by that anonymous "automaton" or "automated subject" (Marx 2012: 127).[23] Yet, capitalism is not a *perpetuum mobile*, it is energised by countless human "machinists" functioning as personifications of specific economic categories (e.g. capital, money, labour power, and consumer power) and following the dominant type of rationality imposed on them by the dominant economic relationships or structures (existing in the form of personal dispositions, attitudes, desires, beliefs, aspirations, etc.).

Generally, capitalists use machines (as means of production) primarily in order to increase productivity, diminish costs, increase profits, and boost competitiveness.

Yet, technology functions – in various ways – also as a tool for controlling and disciplining workers (individually and collectively). For instance, employers take advantage of (physical and mental) labour-saving technology (the result of the information and communication revolution) in order to get rid of rebellious workers, smash their collectives, and drastically reduce the negotiating power of trade unions.[24] Huge unemployment (also caused by macroeconomic stabilisation, e.g. by restrictive monetary and fiscal policies), the growing flexibility of the labour market, the upgraded organisation of production (in line with post-Fordism), neoliberal globalisation (a political response to "the crisis of governability" afflicting both state apparatuses and private capitalist power), and the dismantled welfare state have had an enormous and palpable impact on employed and would-be workers. They have become much more moderate and obedient than during the "revolutionary" period of "the great refusal" when the "spirit of capitalism" was somewhat dying and the intensity of work was diminishing. (Workers did not want to work as they were required to, i.e. as their bosses demanded or expected.) In this manner, we gained yet another proof that fear of unemployment (poverty) and economic uncertainty are probably the most effective emotional "whips" ensuring that loyal and diligent workers gratefully serve private and (to a lesser extent) public masters. Yet, in order to squeeze the maximum efficiency (productive output) from the chosen employees, employers still need additional supervisory measures preventing, as much as possible, idleness, indolence, or, in short, the masters' time (i.e. money) from being wasted (or stolen). Nowadays, machines can also decrease the work of traditional supervisory personnel. For instance, they control (indeed subjugate) workers by imposing a fast tempo of production, reduce the time when they are "doing nothing", register demanded output, score and compare performance levels, monitor the work place with watchful, "panoptic" eyes, and so on. However, the ideal employee is one who enters his or her job with a desire already pre-aligned (or "synchronised") with the master's desire ("algorithm"). One who is already pre-adjusted (pre-programmed) to the company's goals; who functions as a cheerful "*automobile*" (because of his or her "voluntary" motivation to mobilise his or her whole personality in the service of the Other); who is self-watching, self-animated, and self-controlled. Of course, such (fully "involved") workers can enjoy (always more or less limited) autonomy (or "sovereignty") at work, for they are able and willing to eagerly prostitute themselves without despotic, quasi-military methods of hierarchical domination. They can manage and organise their own exploitation by themselves, as perfect slaves (and almost like automated machines or robots).

Concluding remarks on social and class control

We can conclude that sophisticated (monitoring and other) technology still functions as a means for "all too old" purposes, i.e. the exploitation and oppression of structurally subordinated workers. What is new is that capitalist control is becoming increasingly totalitarian, for it aims at subjugating (and reshaping) the post-modern subject to the full: body and soul, external behaviour, and especially thoughts, dispositions, attitudes, desires, tastes, imagination, and (day and night) dreaming. Yet, the (economic, political, and cultural) power of the capitalist class is not based solely or even primarily on its ownership of machinery. Legal structures, defended by "the rule of law" (or *Rechtsstaat*),[25] can be changed overnight, so to speak, as happened, for instance, at the very beginning of the "transition" in ex-socialist countries where the privatisation and denationalisation of state (or "social") property ("wealth") were the cornerstone of the victorious counter-revolution. From the Marxist perspective, it is first of all ideology (much more than repression) that could explain die-hard class domination, at least in fully developed, self-reproducing capitalism (which is able to produce its own structural presuppositions by itself), resembling an "organic system" where each part presupposes all others. Ideology is not the result of intentional manipulation by the economic and political elites or their intellectual apologists, e.g. those responsible for the indoctrination of cadres (particularly lawyers and managers). It springs "naturally" from the basic structures of capitalist society and economy (and from actions that continuously reproduce these structures). Ideological representations – mystifications and fetishes (linked to capital, money, and commodities) – are what spontaneously come to mind (or appear) if one has to function normally in capitalist social relations (or "forms"). As a result of this "religion of everyday life", members of capitalist society live in a "bewitched world turned upside-down" (Marx 1979: 925), i.e. in a reality that is fundamentally non-transparent, regardless of data and information over-abundance. For instance, due to the ideological distortions, one perceives the wage as (usually too low) payment (or a monetary equivalent) for work (or "a job well done"). Profit seems like the achievement of a particular company, obtained in competition with others. Capital, land, and "work" appear to be "productive factors" whose owners are remunerated according to the portion of value that their factor has created or "added". And so on.

Should we suppose, therefore, that combating (or "seeing through") the dominant ideology (the thoughts of superiors and inferiors) would suffice to break or escape from the grip of capitalist power? Well, I am not sure. Namely, even if we are deeply immersed in the ideological mental forms, we still have more than enough information for a revolt, e.g. regarding the astronomic or "sublime" (but entirely unjustified) inequalities in material wealth and income, the irrationality of market competition, and the destruction of nature. However, all this is somehow taken for granted or with apathy, cynicism, resignation, despair, and a sense of powerlessness. ("There is no alternative" but to make constant changes such that nothing essential changes.) My suggestion is that we should take into account also or rather primarily the emotional dimensions of class control, e.g. the notorious fact that capitalist masters manage, indeed extremely successfully, to capture, attract, seduce, shape, and create the desires of the subordinated people in a manner that maintains their domination. That is why economic propaganda, as the chief function of mass media, is so important for the perpetuation of the capitalist status quo. It endlessly celebrates the cult of material wealth (accessible to all, at least "in principle"), the cult of buying and consuming (i.e. destroying) commodities and commercial experiences, and, of course, the cult of the rich, pictured as idols ("holy cows") who should be imitated or, if this is not possible, left in peace, so that they can enjoy their "good lives" without fear. Consequently, we can assume that the main contribution of (post)modern technology in the field of social control (and class war) has to be attributed to its "irresistible" attractiveness in the form of (beautifully designed and increasingly "smart") consumer commodities, e.g. cars, refrigerators, washing machines, television sets, mobile phones, personal computers, etc. In short, apparently, ordinary people like capitalism (much more than socialism)[26] because it looks "so sexy", at least in advertising messages, virtual reality, fiction, and, let us not forget, in the real lives of the rich and famous (the "winners"), and because it offers plenty of cheap "luxury" and colourful trash, *pace* radical critics. It is the imagined heaven of joyful consumption (nowadays in its high-tech version) that justifies the hell of work and maintains its apparatus. It is advisable to strictly separate work and consumption, as we should believe that there is allegedly no causal connection between the rich and the poor, the thin minority of winners and the vast majority of losers. It appears, moreover, that the more people are powerless and confused (e.g. due to the complexity of the globalised world), the more they admire powerful machines. The more artificially constructed stupidity, indeed in its sheer form, invades the postmodern spirit, the more the "smartification" of things advances. The more capitalism seems indestructible (and destructive), the more we are lured into the naturalist thinking according to which, after all, (and in the final analysis), humans are just biological machines run (or programmed) by selfish genetic "algorithms". In summary: Welcome to the overpopulated ship of busy (either happy or sad) "fools", formatted, at least on the surface,[27] according to the model ("instrumental rationality" or calculability) of *homo oeconomicus* (working and consuming man) and increasingly subjugated to machines (e.g. those watching, listening, driving, programming, scoring, and monitoring them), although such machines are supposed to serve and help them. It is a strange world, for it is best designed for and suited to machinery, in relation to which human are predominantly (in work and leisure time) appendices, terminals, and alibis for its being and functioning, although increasingly dispensable and redundant.

Notes

1 For a more precise analysis of the natural and artificial filters (i.e. devices for processing the superabundance of data) used by humans, such as sense organs, consciousness, defence mechanisms, language, culture, and abstract reasoning, see Badiou (1996: 24), Virno (2007: 100–101), Wilson (1990: 236–237), and Žižek (2005: 162).

2 See Arendt (2013: 71), Bauman (2016: 183–191), and Galimberti (2011: 220–224).

3 Algorithms are already widely used in the process of selecting job candidates and hiring employees. Obviously, companies want to find workers who are prepared (or pre-adapted) to serve them with all necessary "human resources", who are eager to mobilise their whole person (body and soul) and energy in accordance with the expectations (i.e. desires) of the master and who will work obediently, loyally, and efficiently. For instance, algorithms grade would-be workers on various indicators of their lifestyle and personality traits, such as conscientiousness, openness to new ideas

(thinking "out of the box"), creativity, agreeableness, neuroticism, extraversion, trustworthiness, ambition, self-control, etc. By using automatic systems (models filtering job candidates), companies want to cut costs, reduce the risk of hiring a bad employee, save money, and gain a competitive advantage.

4 Ironically, in Slovenia, probably the strictest (and consequently) most computerised/algorithmic control is applied with regard to researchers and scholars (see Klepec 2015: 219–220).

5 See Kundera (2008: 134–135).

6 Therefore, it is too naïve to think that in the digitalised era, "algorithms rule the world", regardless of their massive and consequential application in, say, high-frequency trading and managing hedge funds (see Scott 2012: 10; Žižek 2014: 20–23; and Klepec 2015: 216–218). In fact, there is no one in particular who "rules the world" (except perhaps, metaphorically speaking, "King Money").

7 The term "intimacy" is probably more suitable in this context, for it signifies the totality of personal or biographic information that one does not want to share with just anyone. Data of such sort remain an absolute secret or are communicable only to carefully selected persons. On the other hand, "privacy", at least in romantic political philosophy (see Kymlicka 2005: 552–559), refers to a place to which an individual can temporarily withdraw, in order to escape from the social and public parts of the world. Unfortunately, privacy in the sense of an individual's own space and time is too rare a privilege. Often and incorrectly, many identify it with loneliness or with one's family home or family life.

8 It is notable that human beings are also not altogether transparent to themselves. The human soul ("spirit") remains mysterious despite the progress of modern science, for it is, as St Augustine (Avguštin 2003: 204) already observed (long before psychoanalysis and evolutionary psychology), "too tight to embrace itself". It is so not just because of the allegedly powerful unconscious dynamic forces (e.g. repressed emotions and defence mechanisms), but also due to the fact that we are generally not aware of the causes determining our desires ("drives") and actions (see Spinoza 1988: 260). That is why we believe in human freedom of choice, although we do not know why we really want this or that.

9 Scheerer and Hess (1977: 129) assess that the efficiency of post-modern state control has actually declined. Namely, advanced technology is vulnerable. Competent violators can manipulate or even completely neutralise it, for instance by producing false data and obscuring the detection of the correct data. On the other hand, we should also take into consideration the great number, diversification, anonymity, and mobility of human subjects "in need of state control", in addition to the enormous fragmentation of communities, growing social polarisation, and the differentiation of postmodern "societies" into subsystems and subcultures. Therefore, Scheerer and Hess are right in pointing out that the individual human being has never been as "transparent" as in pre-historic communities, early tribal societies, and medieval villages, i.e. in small groups, in which everyone indeed "knew and was able to supervise everybody else".

10 Lea (2002: 159–160) states that the working of post-modern crime control has considerably weakened:

> First there is the weakening identity of many types of crime and offenders and their blurring into wider categories of risk, such that criminalisation starts to lose its taken-for-granted and accepted sets of definitions about what is crime and who are criminal offenders. [...] Second, there is growing reintegration of many forms of crime into the structures of normal social and economic activity. Crime, as social crime, as buying and selling of hard drugs, as money laundering, becomes part of the way the economic system works rather than its breakdown or disruption. [...] Third, the power of criminality to establish forms of governance over

local communities and to neutralise state law enforcement activities also increases under the impact of the increasing wealth and sophistication of criminal enterprise, the undermining of the competence of national states in a global context, together with the undermining of effective governance in many urban areas and in whole regions.

11 Gorz (1999: 14–16) emphasises that this peculiar supra-national power apparatus for expressing and enforcing the "rights" and interests of globalised capital is free of any territoriality, independent from any society and without political constitution.

12 See Hirsch (2014: 85–142) and Supiot (2013: 43–54).

13 Needless to say, plenty of workers like their job or are deeply involved in the work they do for pay. There could be numerous reason for such satisfaction, e.g. the intrinsic nature or indisputable social usefulness of work, a pleasant environment, or emotional atmosphere at the work place, recognition ("love") received from superiors, colleagues, or customers, autonomy (albeit almost always within heteronomy), etc. Of course, there is nothing wrong with workers' having satisfaction with their employment, especially if the jobs in question are socially beneficial. Moreover, we can suppose that many more people would be more happy with their work if working hours were shorter or if they were not constrained (or pressured) by the capitalist imperative of profitable production (see Graeber 2013: 66–68). Yet, various pleasures and joys (or, in other words, objects of desire) associated with one's work situation are usually more or less limited (obviously, one can have a good or even better life beyond one's job). The division of work is also the division of desire (see Lordon 2010: 144–147). On the other hand, apparently many more workers simply become accustomed to work somehow and tolerate their job (the sad fact that others use them as a means to realise their own projects or desires) due to the prospect of purchasing and consuming commercial goods and pre-packaged experiences. Their attitude and motivation regarding their jobs is predominantly instrumental. What really matters for them is the money they need for consumption, as the basis of their social identity, or even their "meaning of life" (see Bocock 1993: 49–52). By functioning in that manner, the individual easily falls in the magic (or rather vicious) circle of the typically post-modern "opiatization of control" (Scheerer & Hess 1977: 119–120), based on the endless creation of needs, wants, or desires. Namely, in order to be able to buy and consume symbolically charged commodities, he or she has to conform to the virtues of the work ethic, and sell his or her own body and soul (their whole person) to the representative of the capitalist system that creates, stimulates, seduces, or shapes (and at least partially or virtually gratifies) his or her desires.

14 There is another paradox. As almost an iron rule, socially and ecologically the most harmful jobs are given the highest financial and, too often, also symbolic rewards. On the other hand, truly useful and even indispensable economic activities are usually poorly remunerated or even denigrated. The world turned upside down? Yes, without a doubt. Yet, unfortunately, the grotesquely unjust distribution of rewards and punishment is just one, although very important, aspect of it (see Galeano 2011: 3–5).

15 See Lessenich (2015: 97–102).

16 See Negri and Hardt (2003: 216–230) and Berardi (2013: 31–32). Gorz (1999. 9–11) emphasises that workers' mass rebellion took many different and radical forms, of which the common denominator was the "refusal to work". There were not just more or less customary strikes. Industrial action also included the rejection of imposed work rhythms, the oppressive (competitive and war-like) logic of productivity and wage differentials, self-ordained reductions in the pace of labour, lengthy occupations of factories, the refusal to transfer negotiating power to the workers' legal representatives (trade unions and leftist political parties), sabotages, etc. It is notable that the outbreak of workers' extreme combativeness was utterly unexpected, for it happened, so to speak, at the peak of the capitalist "golden age", i.e. in the context of almost full

employment, a rising material standard, a solid welfare state, constant economic growth (stimulated by Keynesian monetary and fiscal policies), and "historical" class compromise between organised labour and capital. Yet, it was precisely social and economic security (the absence of the threatening "reserve industrial army") and material welfare that diminished workers' fear of dismissal, enfeebled the authority of capitalist bosses, eroded work discipline, and strengthened class-consciousness and self-confidence (see Bembič 2012: 45–49).

17 There is another aspect (or twist) to the historical irony. Namely, neoliberal capitalism has absorbed quite easily, successfully, and profitably many key cultural demands, values, ideals, or "dreams" of the retrospectively mythologised year of 1968 (see Kobe 2010: 118–124). Post-Fordism has included workers' creativity, flexibility, imagination, emotion, communication, knowledge, inventiveness, learning, self-organisation, autonomy, intellectual abilities, improvisation, ingenuity, and spontaneous co-operation in decentralised production. The "new" boss becomes a manager or rather "coach" who animates and motivates. Employers present work as an opportunity for self-realisation, self-expression, and personal growth. It should no longer be the antithesis of life, but good life as such. Neoliberal propaganda celebrates diversity, deregulation ("it is forbidden to forbid"), individual affirmation (rights and liberty ("free choice"), enjoyment "without limits", the "active" and "sovereign" subject (such as "the state within the state" and "the entrepreneur of one's own human capital"), etc. Moreover, by abolishing wage work and the state, neoliberalism fulfils two basic communist aims, although, unfortunately, in the capitalist social formation (see Virno 2003: 97–98).

18 Foucault (2015: 190) emphasises that absolute poverty (in distinction from the bare "life minimum" or the threshold of biological survival) is also relative. It varies in different societies and cultures. In some places its level is higher than elsewhere. However, poverty of any sort is particularly frustrating in the context of consumerism as "the practical ideology of capitalism" (Bocock 1993: 110) and as "idealist practice" being without limits (Baudrillard, 1988: 24–25). Namely, plenty of postmodern individuals base their self-image, self-worth, and sense of social identities (answers to the question "What am I?", as distinguished from "Who am I?") on their patterns of consumption (not only of goods and experiences, but also of ideas, signs, and symbols associated with the given commodities).

19 See Bryan and Rafferty (2013: 198–200).

20 The increase in productivity (which means that the same consumption of work can produce more products, so that the value of each decreases), due to the use of machine systems, appears ordinary as a result (and the merit) of the productive force of capital. The same also holds for the rise in productivity due to simple co-operation under the command of capitalists and the division of work in manufacturing. The appearance of capital as the power with its own productive force is what Marxist theory dubs "capital fetish", which is not just "false consciousness", a subjective mistake, or a product of manipulation, propaganda, or indoctrination, for its basis is in the capitalist organisation of production. See Lebowitz (2014: 24–29).

21 Technological progress, by increasing productivity, leads to a higher profit rate just for a certain period. After the annulment of the effect of the primary innovation, the differences in profits are levelled again. However, in that process the "organic" structure of capital changes, for less live workforce – in Marxist theory the sole creator of (surplus) value (but not of wealth) – is required for higher productivity. Therefore, after the generalisation of the new technology, the profit rate decreases, not due to diminishing productivity, but because of its growth (see Marx 1973: 270 and Korsika 2013: 55–57).

22 Therefore, the term "social control" might be quite misleading, for members of the bourgeois society are under the firm control of things (see Marx 2012: 60 and Krašovec 2014: 181), and not *vice versa*. It is the movement of commodities (and their market prices) that determines the actions and reaction of capitalists, workers,

and consumers. Neither society nor state (as its political "organ") controls production, exchange, and consumption (e.g. according to autonomously defined needs).

23 In other words, we can detect and blame specific culprits, e.g. bankers, monopolist firms, or financial speculators, only by seeing them as the personalisation of capitalist structures (see Heinrich 2013: 202–209). Should we, therefore, assume that capitalist (social, economic, and legal) structures – i.e. those structures that agents (in various positions and roles) obey and in such manner reproduce in their everyday lives – annul or erase human responsibility and justify altogether the otherwise harmful activities of capitalists (and also workers and consumers)? It would seem so, at least empirically, for one can always say that "everyone would act in the same way if he or she occupied my structural position or role" or that "I am just doing my job" (see Galimberti 2011: 225–227). In that case, capitalist society appears as "organised innocence", if not a "community of saints". The social, economic, human, and ecological damage caused by capitalism is nobody's fault. Shit just happens.

24 See Rifkin (2007: 271–314) and Cohen (2011: 6–13).

25 That means that the material, social, and symbolic privileges of the rich and powerful depend, *in extremis*, on the willingness ("good will") of uniformed and often very badly remunerated men to use their physical force (muscles) and weapons against radical political challengers of the normative status quo. Therefore, in the future, the rich will probably rely much more on the repressive help of robots and an automated defence.

26 Let us take the Slovene example. Retrospectively, the (expired and interred) socialist system seemed to be a workers' paradise (despite its notorious deficiencies), e.g. due to the enviable living standard, economic and social security, leisure time, and tolerable work burdens. Yet, people accepted its political destruction with considerable enthusiasm or even relief. Moreover, it should come as no surprise that the first thing done (in the "transitional" regime of "democracy, human rights, and freedom") was a type of mechanical *lustration*, i.e. the acquisition of new, faster, bigger, safer, Western cars. Logically, that was followed by the accelerated construction of highways ("fast roads") and comfortable "homes" for our holy vehicles. Needless to say, in clear distinction from the "old regime", there is so no serious opposition to the actual capitalist system (in spite of the notable deterioration of the general quality of life). *Amor fati capitalistis*. That is what they wanted, more (and preferably cheap) commodities offered to one's "free choice". That is more than enough to be happy or at least to somehow endure life until its happy end.

27 Beyond that surface, we are still (universally, so to speak) "slaves of affect" (and desires determined from the outside), as Spinoza (1988: 259–262) put it. See also Berardi (2013: 215–217) and Lordon (2010: 30–35).

References

Arendt, H. (2013). *O nasilju* [On violence]. Ljubljana: Krtina.

Arendt, H. (2016). *Condition de l'homme modern*. Paris: Calmann-Lévy.

Avguštin (2003). *Izpovedi* [Confessions]. Celje: Mohorjeva družba.

Badiou, A. (1996). *Etika. Razprava o zavesti o Zlu* [Ethics. A treatise on the consciousness of Evil]. Ljubljana: Društvo za teoretsko psihoanalizo.

Baudrillard, R. (1988). *Selected writings*. Cambridge: Polity Press.

Bauman, Z. (2016). *Postmoderna etika* [Postmodern ethics]. Ljubljana: Znanstvena založba Filozofske fakultete.

Beck, U. (2003). *Kaj je globalizacija? Zmote globalizma – odgovori na globalizacijo* [Was ist Globalisierung? Irrtümer des Globalismus – Antworten auf Globalisierung]. Ljubljana: Krtina.

Bembič, B. (2012). *Kapitalizem v prehodih. Politična in ekonomska zgodovina Zahoda po drugi svetovni vojni* [Capitalism in transitions. A political and economic history of the West after the Second World War]. Ljubljana: Sophia.

Berardi, F. (2013). *Duša na delu* [Il lavoro dell'anima]. Ljubljana: Maska.

Bocock, R. (1995). *Consumption.* London: Routledge.

Brecht, B. (1938). *Poems 1913–1956.* New York: Methuen.

Bryan, D., & Rafferty, M. (2013). *Kapitalizem z derivati: Politična ekonomija finančnih derivatov, kapitala in razrednih odnosov* [Capitalism with derivatives. A political economy of financial derivatives, capital, and class]. Ljubljana: Sophia.

Canfora, L. (2006). *Demokracija. Zgodovina neke ideologije* [La democrazia. Storia di un'ideologia]. Ljubljana: Založba /*cf.

Cohen, D. (2011). *Tri predavanja o postindistrijski družbi* [Trois Leçons sur la société post-industrielle]. Ljubljana: Sophia.

Foucault, M. (2015). *Rojstvo biopolitike. Kurz na Collège de France, 1978–1979* [Naissance de la Biopolitique. Cours au Collège de France (1978–1979)]. Ljubljana: Krtina.

Galeano, E. (2011). *Narobe. Šola narobe sveta* [Patas arriba. La escuela del mundo al revés]. Ljubljana: Sanje.

Galimberti, U. (2011). *Miti našega časa* [I miti del nostro tempo]. Ljubljana: Modrijan.

Gorz, A. (1999). *Reclaiming work. Beyond the wage-based society.* Cambridge: Polity Press.

Graeber, D. (2013). *Fragmenti anarhistične antropologije* [Fragments of Anarchist Anthropology]. Ljubljana: Založba /*cf.

Heinrich, M. (2013). *Kritika politične ekonomije: Uvod* [Kritik der politischen Ökonomie: Eine Einführung]. Ljubljana: Sophia.

Hirsch, J. (2014). *Gospostvo, hegemonija in politične alternative* [Herrschaft, Hegemonie und politische Alternativen]. Ljubljana: Sophia.

Klepec. P. (2015). Kaj je strojno podjarmljanje? [What is machine-generated subjugation?]. *Problemi, LIII*(7–8), 163–230.

Kobe, Z. (2010). Tri teze o postfordizmu. In G. Kirn (Ed.), *Postfordizem: razprave o sodobnem kapitalizmu* [Post-Fordism: Studies on contemporary capitalism] (pp. 113–132). Ljubljana: Mirovni inštitut, Inštitut za sodobne družbene in politične študije.

Korsika, A. (2013). Kapital, profit in kriza [Capital, profit, and crisis]. *Borec, LXV*(694–697), 26–47.

Krašovec, P. (2014). Razpoke v socialističnem trikotniku (Spremna beseda) [The schisms in the socialist triangle (commentary)]. In M. A. Lebowitz, *Socialistična Aleternativa* [Socialist alternative] (pp. 165–187). Ljubljana: Sophia.

Kundera, M. (2008). *Zastor: esej v sedmih delih* [Le rideau: essai en sept parties]. Ljubljana: Modrijan.

Kymlicka, W. (2005). *Sodobna politična filozofija. Uvod* [Contemporary political philosophy. An introduction]. Ljubljana: Krtina.

Lea, J. (2002). *Crime & modernity. Continuities in left realist criminology.* London: Sage.

Lebowitz, M. A. (2014). *Socialistična alternativa: Resnični človekov razvoj* [The socialist alternative: Real human development]. Ljubljana: Sophia.

Lessenich, S. (2015). *Ponovno izumljanje socialnega: Socialna država v prožnem kapitalizmu* [Die Neuerfindungen des Sozialen: Der Sozialstaat im flexiblen Kapitalismus]. Ljubljana: Krtina.

Lordon, F. (2010). *Capitalisme, désir et servitude.* Paris: La fabrique éditions.

Marx, K. (1973). *Kapital,* 1. knjiga [Capital, first book]. Ljubljana: Cankarjeva založba.

Marx, K. (2012). *Kapital,* 3. knjiga [Capital, third book]. Ljubljana: Sophia.

Negri, A., & Hardt, M. (2003). *Imperij* [Empire]. Ljubljana: Študentska založba.

Rifkin, J. (2007). *Konec dela. Zaton svetovne delavske sile in nastop posttržne dobe* [The end of work. The decline of the global labor force and the dawn of the post-market era]. Ljubljana: Krtina.

Scheerer, S., & Hess, H. (1997). Social control: A defence and reformulation. In R. Bergalli, & C. Sumner (Eds.), *Social control and political order. European perspectives at the end of the century* (pp. 96–130). London: Sage.

Scott, J. H. (2012). *Automate this. How algorithms came to rule our world.* London: Penguin.

Spinoza, B. (1988). *Etika* [Ethics]. Ljubljana: Slovenska matica.

Supiot, A. (2013). *Duh Filadelfije. Socialna pravičnost proti totalitarnemu trgu* [L'esprit de Philadelphie. La justice sociale face au marché total]. Ljubljana: Založba /*cf.

Virno, P. (2003). *Slovnica množtva. K analizi oblik sodobnega življenja* [Grammatica della moltitudine. Per una analisi delle forme di vita contemporanea]. Ljubljana: Krtina.

Virno, P. (2007). Družbene vede in "človeška narava". Govorna sposobnost, biološka stalnica, proizvodni odnosi [Scienze sociali e "natura umana". Facoltà di linguaggio, invariante biologico, rapporti di produzione]. In M. Foucault, N. Chomsky, & P. Virno, *Človeška narava in zgodovina* [Human nature and history] (pp. 63–165). Ljubljana: Krtina.

Wilson, C. (1990). *Psihologija ubistva* [The psychology of murder]. Niš: Gradina.

Žižek, S. (2005). *Kako biti nihče* [How to be nobody]. Ljubljana: Društvo za teoretsko psihoanalizo.

Žižek, S. (2014). Kriza, kakšna kriza? [Crisis, what crisis?]. *Problemi, XLIII*(7–8), 5–30.

Part III

Automated policing

5 Data collection without limits

Automated policing and the politics of framelessness

Mark Andrejevic

The police equipment company formerly known as Taser announced in 2017 that it was offering free body cameras – almost a half-million of them –to police officers across the country. In exchange, the company wanted access to all of the captured video in order to train its emerging artificial intelligence systems how to process police video automatically (Coldewey, 2017). The strategy was a familiar one in the tech sector: lead with a loss in order to capture a data trove. Google, for example, pioneered this strategy with email: when other email companies were trying to hook users with free email accounts in order to get them to pay for more storage, Google offered large amounts of storage free in return for access to the data. "You can store as much email as you want on our servers," was the implicit offer, "If you just let us mine it in order to learn more about you and to target ads more effectively." And so an economic model for social media was born. The policing company, Axon, is anticipating one of the pressing imperatives of the twenty-first century: the ability to process large amounts of data automatically – without human assistance – in order to deal with the huge volumes of information being captured by security cameras, body cameras, and cell phones. In an era of information glut, the challenge is not getting the data, it's making sense of it and putting it to use. Taser was so enthralled by the prospects of this new business model that it changed its name to Axon (a biological reference to communicative infrastructure) in anticipation of its announcement. The company's original goal, according to its founder, was, "to make the bullet obsolete" (Coldewey, 2017). Now the company is joining the information processing bandwagon, as policing makes the datalogical turn. The promise is not just that information will make policing more efficient, but that information might contribute to the goal of pre-emption: intervening in real time before a crime can take place. Policing, the name change suggested, is now more about information than weaponry: or, perhaps, the weaponization of data.

The progressive automation of sight (and foresight)

In the emerging sensor society, the information-processing bottleneck is the human being. Data collection has become relatively easy and cheap, but many forms of data processing still rely on direct human engagement. Millions of cops with cameras – like roaming drones – can capture a huge trove of information, but making sense of it requires even more time than capturing it. As Axon's CEO put it, "The real challenge is that this is a data problem much more than a hardware problem.... This is a massive amount of information compared to any IT department you would find in a police agency" (Ng, 2017). With 95 percent of police departments across the country predicting that they will equip their officers with body cameras, the trove of video data promises to be tremendous (Ng, 2017). Add to that the video from dash cams and security cameras, and the need for automated video processing becomes pressingly obvious. Moreover, as the video database becomes increasingly comprehensive, the goal is not simply to process captured information but to use if for predictive purposes: to guide future action. As Paul Virilio put it in his anticipatory remarks on the rise of automated vision (what he referred to later as the "vision machine"): "seeing and foreseeing therefore tend to merge" (1989, 4). Augmented vision and automated sensing tend to coalesce around the goal of pre-emption, as suggested by the logic of the body cam, which is described as having the potential for a deterrent effect on both officer and suspect (though the promise of deterrence, in this register, relies on a disciplinary logic which is, as we shall see, called into question by the advent of so-called predictive policing). As the CEO of a company that is developing camera-equipped smart glasses for police officers (to automatically capture number plates, faces, and so on) put it, "Live-streaming and license plate recognition is only the beginning.... Our situational awareness and anomaly detection capabilities will allow organizations in the public sector to react faster to real-time events" (*Business Wire*, 2016). Automated anomaly detection pushes one step beyond the augmented forms of detection associated with the development of "cortically coupled vision" – a system that takes advantage of the cognitive lag between when a viewer's brain registers an anomaly and when the recognition of that anomaly becomes conscious. In cortically coupled vision systems, an EEG device attached to a human observer skips the process of conscious recognition, shortening response time. When the brain responds to the anomaly, there is no need to wait for the viewer to realize what happened: instead, the brain signal is directly relayed to a computer. According to one news account, "The Army is interested in using such a mind-machine interface to help soldiers navigate dangerous terrain" (Daley et al., 2011). A device coupled to the driver's brain (via EEG) would register an anomaly response before the driver is consciously aware of it: "His C3Vision headgear would register the brain waves associated with the suspicious object and inject them into the vehicle's driving system. When the system sees other things out there that look similar, it would automatically evade them" (Daley et al., 2011). The brain is redoubled as a sensor that can dispense with the subject's conscious recognition.

The logical next step – as with a range of informated technologies these days – is to subtract the human being entirely, training machines to recognize the patterns that triggered the human response. As the system's inventor puts it, "The system latches on to individual perceptions and trains the computer to know what the user means by interesting [or alarming]" (Daley, Piore, Lerner, & Svoboda, 2011). Once the human is subtracted from the system, the vision machine comes to rely on what Trevor Paglen describes as "operational images": "Namely images made by machines for other machines" (2014, 2). And when this happens, of course, the image is no longer necessarily "visual": "machines rarely even bother making the meat-eye interpretable versions of their operational images.... There's really no point. Meat-eyes are far too inefficient to see what's going on anyway" (3). The expansion of vision is associated with the surpassing of the viewing subject – a dynamic that lies at the heart of contemporary logics of data-driven pre-emption.

The surpassing of human ("meat-eye") vision marks a moment of convergence: *whatever* a sensor captures can participate in the data image that is used to identify, predict, and pre-empt. Video becomes simply one more possible input. The equipping of police offers with body cams may have been prompted by calls for accountability, but it fits within the accelerating logic of comprehensive information capture associated with the rise of data driven policing. If the goal is not simply recollection or accountability (to capture and record an interaction), but to envision the possibility of automated anomaly detection in order to predict and pre-empt criminal activity, the broadest possible sweep of data collection is required. This imperative of total information awareness is built into the rallying cry for big data mining, epitomized by the CIA's goal, "to collect everything and hold on to it forever" (Sledge, 2012). The imperative is a result of the speculative character of data mining techniques designed to uncover unanticipated and unintuitable correlations: if the goal is to find completely unexpected patterns, there is (by definition) no a priori limit on what information might end up being useful. Thus, in the realm of predictive policing, it's becoming common to push beyond the customary forms of historical data that guided policing practices: past patterns of criminal activity. Now data mining operations are getting more creative in the types of information they incorporate into their predictive analytics. A company called HunchLab includes factors such as "population density; census data; the locations of bars, churches, schools, and transportation hubs; schedules for home games – even moon phases" (Chammah, 2016, para. 7). Although some correlations are to be expected, "Others are more mysterious: rates of aggravated assault in Chicago have decreased on windier days, while cars in Philadelphia were stolen more often when parked near schools" (Chammah, 2016). The impetus to push beyond historical crime data is reinforced by critiques of the historical biases in law enforcement: more crimes can be documented in areas that are more heavily policed. As one account put it, "Systems that incorporate these sorts of statistics may not account for the inaccuracies reflected in historical data, leading to a cycle of self-fulfilling prophecies" (Koepke, 2016). The automation of data processing, then, pushes in the direction of what might be described as a form of "framelessness": there is no limit on the type or amount of data that may be relevant to predicting either individual behavior or spatial patterns of crime. Frames narrow down the scope of potential relevance of information and thus of data collection itself. "Frameless" data collection does not rule out anything in advance.

Beyond the frame

We live in a time when our media technologies stage for us an emerging aesthetics of framelessness: virtual reality, augmented reality and 360-degree cameras offer the spectacle of totalization – the "complete" capture of a reality, whether constructed or observed. As always, there are historical precedents. In particular the fascination with panoramas speak to an ongoing fascination with what Martin and Moar (2017) describe as "immersion." They trace this fascination back to the Great Frieze in Pompeii's Villa dei Misteri and then to the large panoramas of the eighteenth and nineteenth centuries. One of the defining experiences they recount, drawing on testimonials of viewers – even present day ones, is the experience of framelessness: "Nowhere is it possible for the eye to shift 'outside the frame' and compare the artistic illusion with the real surroundings" (Martin and Moar, 2017; 2). The fascination of such an experience is one of transport – an experience not unrelated to the simultaneous development of the railroad, which inaugurated the time-space compression of modernity. The panorama promised a kind of teleportation: the ability to travel the world without leaving home (a promise later picked up on by Microsoft's "Where Do You Want To Go Today" campaign for its Windows operating system). In the contemporary context, the promise of immersion is about something slightly different, not so much transport as information access: the totalization of representation associated with simulation.

If the railroad ushered in the fantasy of high-speed transport, the internet refreshed the promise of total information access: an immaterial, digitally datafied form of time-space compression. Against the background of such a promise, information incompleteness emerges as an increasingly pressing obstacle to overcome. Once upon a time, the defining dimension of media critique was scarcity: the inability to obtain complete information, the partial and incomplete picture crafted by media gatekeepers and enforced by the economic and technological entry barriers to the realms of media production and distribution. In an era of relative information scarcity, the field of media studies took up the notion of the frame to describe the process whereby media gatekeepers decide what counts as news and how a particular story is crafted according to unarticulated, unstated, and sometimes unconscious preconceptions. The media critic Todd Gitlin describes media frames as, "persistent patterns of cognition, interpretation, and presentation, of selection, emphasis, and exclusion, by which symbol-handlers routinely organize discourse, whether verbal or visual" (Gitlin 1980; 7). In concrete terms, the notion of a "frame" is used to examine the perspective from which a story is told: does a news item about inflation focus on the Wall-Street impact or the consumer impact; does a story about a strike focus on the inconvenience to consumers or the hardships faced by workers? The convention of objectivity works to dissimulate the existence of a frame by implying that the selection of facts and their presentation in a story takes place neutrally: that the world is being presented simply "as it is."

The notion of objectivity has been the target of substantive critique in the mass media era, particularly when it came to elucidating the ways in which dominant narratives reflected and reproduced the priorities of political and economic elites. It has become a truism that all news stories are partial – in both senses of the word – and this truism has been transformed from a tool of progressive critique into a generalized savviness and debunkery that can just as easily serve those in power as their critics. Bruno Latour lamented this development in his essay on the fate of critique: "Has knowledge-slash-power been co-opted of late by the National Security Agency?" (2004, 228). The reason the "fake news" cry was so readily co-opted by those on the political right was because it had long been their charge against the "mainstream" media: that it was biased, incomplete, and manipulative. To put it somewhat differently, the driving force behind the contemporary critique of representation is that it falls short of framelessness: it can never capture the full reality of what is going on and is thus always subject to bias inaccuracy, incompleteness. As critique turns back upon itself, it becomes increasingly reactionary. This progression is a familiar one: the recognition of the necessary existence of a frame is mobilized as a delegitimizing tool that can be directed against any possible narrative account. The result is the proliferation of increasingly convoluted conspiracy theories – facilitated in part by their political mainstreaming during the ascendency of Donald Trump and the media outlets associated with his supporters (Alex Jones's *InfoWars* and *Breitbart*, for example). These are outlets that trade on their readers' awareness of the room opened up for doubt and suspicion by the necessary incompleteness of all (mainstream) media accounts. The very fact that they pretend to be unbiased and impartial (therefore, for all practical purposes, complete) renders them suspect.

The rejoinder to the pathologies of the media frame, in the current media moment, is the promise of totality, taken up on the one hand by the promise of big data and on the other by its representational equivalent: virtual reality (which offers the prospect of reduplicating reality in its entirety). It is suggestive, in this regard, that journalists (like Nonny De La Pena, for example – see Garling, 2015) are suggesting that virtual reality might be a more effective way of communicating the news than conventional narrative reporting. Instead of telling people a story, virtual reality can place them in a situation and let them feel like they are experiencing it directly, on their own terms. The promise, of course, is false, but the representational form dispenses with the conventions of the frame – of the video camera (which raises the question about what is left out) or even the delimited narrative (which can never describe reality in its entirety). Similarly, Facebook founder and CEO Mark Zuckerberg (whose company now owns Oculus Rift, the storied manufacturer of VR headsets) envisions the possibility of directly communicating experience via virtual reality:

> I think you're going to be able to capture a thought. What you're thinking or feeling, in its kind of ideal and perfect form in your head and be able to share that with the world in a format where they can get that.
>
> (Dewey, 2016)

In both cases, narrative is displaced by immersion: the fantasy of immersion becomes not one of transport but of communication – that is, the prospect of bypassing the partial character of representation in order to present a world in its entirety. The goal of framelessness implies both the possibility of total information capture – that is, the multiplication of sensors across the landscape – and that of total information representation: in the database (which we can't "see") and VR (which we can).

The fact that total information capture is impossible does not prevent it from being embraced as an ideal. We can already identify technologies that move in this direction, such as the development of so-called "smart dust" sensors that, "float in the air throughout the entire city and track movement, biometric indicators, temperature change, and chemical composition of everything in their city" (Rowinski, 2013). Perhaps the most obvious example is that of the so-called "internet of things," which promises to render much of the object world interactive (in the sense of being able to capture and communicate information) by providing it with a "smart" overlay. We find traces of this overlay cropping up in various guises devised by a range of corporate players. Amazon Go stores, for example, anticipate a world in which stores track individual users as they move through the aisles, using cameras, Wi-Fi, microphones, and a range of other sensors to monitor and analyze every movement in detail. Just as Web browsers can track the actions of individuals, so-called "smart" spaces promise increasingly comprehensive forms of data collection and automated response. Since we always inhabit physical space, and only sometimes "virtual" space, the possibilities for information collection become increasingly comprehensive as interactivity colonizes our lived environment. We might describe ubiquitous computing as "environmental interactivity," as long as we understand this primarily as a *passive* form of interactivity: the spaces through which we move and the objects with which we interact will create a detailed database about us, without any additional effort on our part.

There is a neat fit between the expanding goal of information capture for the purposes of prediction (and pre-emption) and the "becoming smart" of our increasingly interactive spaces and things. If the goal of securitization is to capture everything and hang on to it forever, the interface that supports this goal is the internet of things. It will likely become increasingly common to hear about police collecting data from digital personal assistants like Alexa not just as evidence, but perhaps eventually for the purposes of profiling and predicting. The consequences for policing of what might be described as data collection without limits – or frameless data collection – are multiple, but this chapter will focus on the following: the rise of post-disciplinary logics of securitization, the rise of "environmental" surveillance (in the sense of smart spaces and objects), reconfigured logics of targeting, the shift from prevention to pre-emption; and the displacement of explanation by correlation.

Post-disciplinarity and the fate of the subject

Disciplinary societies, according to the work of Michel Foucault (2012), rely on logics of subjectification that underwrite the efficiency of panoptic forms of control. Thus, Foucault can argue – still within the horizon of "framed" surveillance – that panopticism does not require "total" monitoring. As Foucault puts it, in the disciplinary prison,

> it is at once too much and too little that the prisoner should be constantly observed by an inspector: too little, for what matters is that he knows himself to be observed; too much, because he has no need in fact of being so.
>
> (2012, 201)

The disciplinary model relies on monitored subjects internalizing the imperatives of the watchers. Because they discipline themselves (with the help of outside coercion), they do not need to be watched all the time, and perhaps eventually not at all. As Elayne Rapping (2004) notes, the disciplinary model presupposes the potential for subjectification – that is, it envisions subjects who will respond to the connection between surveillance, transgression, and punishment according to a predetermined and uniform version of rationality. As might be expected, however, the logic of framelessness dispenses with the disciplinable subject (and even the process of subjectification) as such (not least because the subject is a kind of frame that determines the partial and perspectival character of experience).

Control processes shift in the wake of the subjective dissolution. Elayne Rapping (2004), for example, argues that the notion of the disciplinary subject breaks down in contemporary policing practices and she draws on the portrayals of criminality in the reality TV show *Cops* to highlight this shift. Whereas traditional crime dramas, she argues, portray subjects who remain legible within the horizon of discipline (they may be deviant, but there is some kind of explanation for their conduct – one that is left to the detective to unearth), *Cops* portrays the dissolution of the disciplinary subject. On *Cops*, criminals are portrayed as "incorrigibly 'other' and 'alien;' incapable of internalizing or abiding by the norms and values of a liberal democracy, for they are far too irrational, uncontrollable, and inscrutable for such measures to be effective" (Rapping, 2004, p. 227). The criminals portrayed on *Cops* are not figures who participate in the rational calculus inspired by the threat of surveillance and the specter of punishment: they come to represent the resurgence of the category of the undisciplinable subject. Such portrayals align with the reconfiguration of risk in the post-9/11 era and the subsequent rereading of criminality through the lens of terrorism. The paradigmatic representative of a non-narrativizable, ubiquitous threat, she argues, is the specter of the terrorist: "Terrorists are irrational, inscrutable, and inherently violent.... And they cannot be 'reformed' or 'rehabilitated' according to traditional correctional methods because they neither recognize nor respect the codes to which such measures apply" (p. 225). Or at least that is the prevailing narrative characteristic of the post-political response to terrorism. To revisit Foucault's formulation, for the figure of the terrorist, subjection to panoptic surveillance is at once too much and not enough – too much because they do not need to be *told* they are being monitored (this would not have the desired disciplinary effect) and not enough because they must be monitored *all the time*, as comprehensively as possible. Post-panopticism invokes the goal of framelessness: that of total information capture. All spaces and people must be monitored all the time. If there was something "light" about the apparatus of the Panopticon (Foucault observes that Bentham was, "surprised that panoptic institutions could be so light" (2012, 202)), the apparatus becomes somewhat heavier in the post-panoptic moment. The entire infrastructure becomes redoubled as an information gathering system: our appliances, our cars, our "digital assistants," our schools and shops all become automated data collection systems.

The "becoming environmental" of surveillance

The interactive infrastructure provides information not just for marketing but also for security and policing. The development of so-called predictive policing has been interpreted as having two dimensions: the spatial (when and where crime is likely to take place) and the personal (who is likely to commit a crime) (Koepke, 2016). For the moment – perhaps because of privacy concerns – the focus has been largely upon the former (with the notable exceptions of programs like Chicago's "Heat List," which targets "high risk" individuals with messages that they are being watched – marking the persistence of disciplinary approaches). As the co-creator of one of the more well-known predictive policing platforms, PredPol, puts it: "This is not about predicting the behavior of a specific individual.... It's about predicting the risk of certain types of crimes in time and space" (Beam, 2011). In principle, however, the distinction between personal and spatial surveillance is a tenuous and temporary one. Predictive policing starts off with historical data – which tend to be location specific: the traditional "pins on the map" approach. However, as both Rapping's work and historical trends indicate, contemporary policing is increasingly shaped by the imperatives of counter-insurgency. (A trend that, as Wall, 2013, notes, has deep historical roots in policing, both domestic and colonial.) In concrete terms, this means that the entire population is approached as a locus of risk (whether of insurgency or criminality). The risk calculation becomes a probabilistic one: where and when will a member of the population emerge as perpetrator? Criminality is treated as endemic in the same way that insurgency remains a constant possibility: the ideal of total pacification is replaced by that of perfect pre-emption: since crime will "never go away" and irrationality is always with us, according to this account, the policing process is a perpetual one and surveillance must be ubiquitous and continuous. Elaborating on the imperative of total information collection, Ben Anderson describes the logic of counter-insurgency as follows:

> because popular support is never definitively achieved, attempts to produce and harness it must be continuous. They must also extend throughout life without limit because everything has the potential to initiate the becoming

> (counter)insurgent of the population.... Dangerousness exists as a potential distributed everywhere and conditioned by anything and everything.
>
> (2011, 224)

This formulation of surveillance throughout life without limits is another way of approaching the notion of framelessness, and it applies equally to the process of policing described by Rapping.

We don't have to look far to locate the "becoming environmental" of surveillance. Barack Obama's director of National Intelligence said in 2016 that the internet of things might be used for, "identification, surveillance, monitoring, location tracking, and targeting for recruitment" (Ackerman & Thielman, 2016). The formulation compactly invokes several dimensions of framelessness: the entire environment becomes an information capture device (there is no limit on the designated times and spaces of monitoring: everything is fair game, all the time); everyone is subject to the monitoring gaze; and even the function of surveillance is not limited to a particular frame or context. (Monitoring can be used to identify suspects and allies alike, as well as consumers, prospective employees, etc.) The capture of so much information for so many possible purposes renders impossible the notion that it could be managed by individuals: capturing everything about everyone necessitates the development of post-subjective perception and the rise of the operational "image." The notion of the image here extends beyond the visual – from the perspective of digital machine perception, visual data merges with all the other bits gathered by the interactive apparatus. The operational image is post-visual and perhaps post-representational: it combines the historical, the visual, the demographic, and more into patterns that can be mined for correlations.

The target is the population

The aptly named policing software system "Beware" promises to provide first responders with real time threat assessment profiles of the entire population. It is based on the capture of as much data as possible about every address in the United States. One news account describes the software in action, "While officers raced to a recent 911 call about a man threatening his ex-girlfriend, a police operator in headquarters consulted software that scored the suspect's potential for violence the way a bank might run a credit report" (Jouvenal, 2016). The comparison is a suggestive one: credit ratings aspire to universality: the goal is that there is one available for everyone, in order to facilitate economic transactions. Something similar is happening with threat assessment scores: everyone in the US now has one, thanks to systems like Beware. The scoring system pushes beyond a disciplinary system that separates deviants from non-deviants. It is an actuarial system that places everyone along the spectrum of risk and threat, with the understanding that anyone might become subject to a police request for a threat assessment score at some point. If anyone can pose a potential risk, then the surveillance imperative is to gather as much information as possible about everyone. As Anderson puts it, "Dangerousness exists as a potential distributed everywhere and conditioned by anything and everything" (2014, 224).

Such an approach supplements strategies of individual targeting with those of population level monitoring. The point is no longer to start by isolating target individuals, but rather to gather information about everyone in order to identify the patterns that help predict and pre-empt the potential to "become criminal" as it arises. Of course the goal is also to manage risk in confrontations by pre-empting escalation. Thus, the Beware application relies on scraping as much data as possible about as many people as possible: "The program scoured billions of data points, including arrest reports, property records, commercial databases, deep Web searches and … social-media postings" (Jouvenal, 2016).

The goal of wholesale data collection was given an extra boost in the United States with the 2017 decision by Congress to roll back privacy regulations that would have given users greater control over the sale of information collected about them by their Internet Service Providers (ISPs). As more and more devices go online – from smart refrigerators to personal digital assistants – ISPs will have access to a growing stream of data and the new rules provide them with increasing flexibility to market it to services like Beware and other commercial predictive policing systems, further extending the data collection process "throughout life without limit" (Anderson, 2011; 224). As one news account put it, "The 'big data' that has transformed marketing and other industries has now come to law enforcement" (Jouvenal, 2016).

From prevention to pre-emption

The displacement of the disciplinary model (which is not the same as its wholesale *re*placement) described by Rapping (2004) results in an emphasis on real-time pre-emption rather than long-term prevention. The latter relies upon processes of interpretation and explanation that posit disciplinary subjects amenable to knowledge-based reform practices reliant upon the vocabulary of "deviance, delinquency, reform, and rehabilitation" (Rapping, 2004, 228) embraced by twentieth-century discourses of sociology and criminology. Pre-emption, by contrast, relies on data driven decision-making processes that prioritize correlation over explanation, and thus align themselves with the "end of theory" mindset elaborated by Chris Anderson:

> Out with every theory of human behavior, from linguistics to sociology. Forget taxonomy, ontology, and psychology. Who knows why people do what they do? The point is they do it, and we can track and measure it with unprecedented fidelity. With enough data, the numbers speak for themselves.
>
> (2008)

The perceived virtue of such a approach is that it promises to transcend the deadlock of deliberation associated with interpretation and deliberation. We need not, as a society, come to an agreement about the possible underlying *causes* of crime (social conditions, personal pathologies, genetics, the decline of the welfare state, etc.) to deploy pre-emptive strategies that address its manifestations (rather than its underlying conditions).

Predictive policing frames itself in terms of real-time intervention: police officers showing up just in time to prevent a crime, or patrols sent to a location in order to deter impending violence. The data may be there to suggest long term correlations: between, say, literacy and crime rates – but addressing such issues requires political intervention and long-term policy making that fail to address the repeatedly emphasized *urgency* of the contemporary threat. As the philosopher Gregoire Chamayou (2015) observes in his work on counter-terrorism: "Within the categories of policing, political analysis dissolves" (69). This dissolution is becoming increasingly familiar in the portrayal of criminality traced by Rapping – one in which the criminal inhabits the domain of the irrational, implacable, and undisciplinable. The assimilation of criminality to terrorism dispenses with the hope that policies other than direct pre-emption might matter. The attempt, for example, to suggest that there might be underlying socio-political causes (and solutions) for terrorism is treated as a form of terrorist complicity and ridiculed as vestigial liberal do-gooderism that simple-mindedly advocates "jobs for Jihadis" (Greenberg, 2015). The technological and political solutions become increasingly incompatible as the former become a way of denigrating the latter.

This incompatibility between technological and political solutions has consequences for policy discussions and the public's understanding of the role of police surveillance in the database era. Whereas once upon a time, those concerned about crime rates might have inquired into the underlying social causes of violent or criminal behavior, the question now shifts in the direction of "Do we have data to predict and preempt it?" And the answer is a familiar one: we can always use more data. As recently as the 1990s, urban crime could be approached, in regulatory terms, as a social issue that raised the question of how conditions of at-risk youth might be addressed by social programs. The federal Local Law Enforcement Block Grant program, which once provided support for community crime prevention, saw its funding drop precipitously in the post-9/11 era and was replaced completely in 2004 to make way for the Federal Justice Assistance Grant program, which focuses on equipment and training (Bauer, 2004). Midnight basketball programs gave way to the database and algorithm (as well as enhanced technologies for surveillance). In the schools and on the streets the discourses of community engagement are displaced by the promise of the technological fix: more biometrics, more tracking, more social media analysis. As one article on security surveillance in the schools observes, authorities turn to these devices and technologies, "rather than building community relationships aimed at getting to know family members, for security" (Nguyen, 2015, 5).

The future of modulation

The contemporary resurgent fascination with the promise of virtual reality (*this time the hype is justified*, we are told) invokes an additional aspect of immersion:

the prospect of its complete customization. What differentiates the Oculus Rift from the panoramas of yore is its potential for modulation in the sense invoked by Gilles Deleuze in his essay on control societies: "controls are a modulation, like a self-deforming cast that will continuously change from one moment to the other, or like a sieve whose mesh will transmute from point to point" (1992, 4). Panoramas, like the movies of the mid-twentieth century, were collective spectacles. Other early immersive technologies, like stereopticons may have relied upon individual viewing, but they were not customizable: the picture remained the same for each viewer. Virtual reality anticipates the possibility of fully customizable information environments presaged by the filter bubble or the custom news feed. Frameless representation, in this regard, promises the possibility not only of totality (the viewer feels as though the picture is unframed – and thus complete, unlike a photograph or conventional video) but also of a *unique* totality for each viewer, one that can adjust in real time to the viewer's responses. The type of environmental governance described by Anderson relies not just upon environmental surveillance (the full penetration of monitoring into the rhythms of daily life), but also on environmental modulation: the customized adjustment of individual information environments in accordance with the logic of pre-emption. Perhaps this is the fantasy that lies at the heart of "frameless" systems of control: not only total information awareness, but also perfect information modulation. In political terms, this is the promise of a company like Cambridge Analytica, which claims to know so much about individual voters that it can manipulate them through custom-targeted information campaigns (Cadwalladr, 2017). From the perspective of policing, the logical extension of strategies of pre-emption is to intervene through processes of environmental modulation that can become at once increasingly comprehensive and targeted. As Slavoj Zizek has observed, the prospect of complete freedom (from the constraints of reality) envisioned by virtual reality goes hand-in-hand with that of total control (by those who operate the platform): "on the one hand, reduction of reality to a virtual domain regulated by arbitrary rules that can be suspended; on the other hand, the concealed truth of this freedom, the reduction of the subject to an utter instrumentalized passivity" (1999, 26).

Subjective obliteration

By way of summation – and to make a somewhat broader point about emerging logics of data-driven pre-emption – we might trace the underlying post-subjective logic of automated control. The fantasy of automation and that of totalization merge: when the information collection process becomes increasingly comprehensive it creates a cascading logic of automation: automated data acquisition generates huge information troves that require automated processing and tend toward automated response. The prospect of framelessness associated with total information collection envisions the collapse of those gaps that are characteristic of both the subject (necessarily delimited) and of language. The target of framelessness is a perceived lack – and lack is at the basis of subjectivity (and language), so, it should come as no surprise at this point that the target of framelessness becomes the subject itself. In logical terms the condition of the existence of a distinct and unique subject is finitude. (The stirrings of the subject emerge when an individual realizes that it is, literally, "not-all" – that it is distinct from its surrounding world, dependent on it, but unable to control it completely.) Similarly, the condition of language is lack (the representation of what is not there – as well as the gap between what can be said and what is meant). Language is inherently perspectival, delimited, partial – in short subjective. This lack is precisely what the critique of narrative and mediation holds against them: that they cannot provide us with complete presence. Think of the conspiracy theorists who will never believe that, for example, Barack Obama was born in the United States: the only proof that could possibly overcome their objections would be to have *been there* for the event. No representation could ever attain the impossible fullness they require as proof. One of the goals of virtual reality is to make it seem – as seamlessly as possible – that *we are there*. Of course, it remains a form of mediation – one whose goal would be the obliteration of the experience of mediation itself: the psychotic collapse of the representation into reality itself. Zizek describes this collapse as one that eradicates "the distance (between 'things' and 'words') which opens up the space for … symbolic engagement" (1996, 196). What I have described as the dissolution of politics and the post-narrative imperative of data-driven pre-emption are both ways of describing this eclipse of symbolic engagement.

The goal of total information capture is the perfection of prediction and the pre-emption of desire. In the realm of consumption, this logic manifests itself in the form of the womblike promise of addressing needs before they arise. The automation of commerce envisions the fulfillment of the marketer's promise to "know what you want before you want it" (and to deliver it in real time). In the realms of policing and security, the promise of total information awareness is related: to intervene and modulate the environment before a subject is able to realize a desire to do harm. The process of total information capture and total environmental control is pre-disciplinary, it does not rely upon the process of subjectification, but envisions intervention at another level – one that short circuits subjective forms of control and decision-making. At both ends of the process, then, the logic of framelessness anticipates the collapse of the subject: through its pre-emption on the one hand, and its surpassing, in the form of automated technological systems, on the other.

References

Ackerman, Spencer, & Thielman, Sam. (2016, February 9). U.S. intelligence chief: We might use the Internet of things to spy on you. *Guardian*. Retrieved from www.theguardian.com/technology/2016/feb/09/internet-of-things-smart-home-devices-government-surveillance-james-clapper.

Anderson, Ben. (2011). Facing the future enemy: US counterinsurgency doctrine and the pre-insurgent. *Theory, Culture & Society*, *28*(7–8), 216–240.

Anderson, Chris. (2008, June 23). The end of theory: The data deluge makes the scientific method obsolete, *Wired Magazine*. Retrieved online March 22, 2017 at: www.wired.com/science/discoveries/magazine/16-07/pb_theory.

Bauer, Lynn. (2004). *Local law enforcement block grant program, 1996–2004 (Technical report)*. Washington, DC: Bureau of Justice Statistics.

Beam, Christopher. (2011, January 24) Time cops. *Slate*. Retrieved online March 22, 2017 at: www.slate.com/articles/news_and_politics/crime/2011/01/time_cops.html.

Business Wire. (2016, November 25). Chinese traffic police use smartglasses to read license plates and test situational awareness. Retrieved online March 22, 2017 at: www.businesswire.com/news/home/20161125005300/en/Chinese-Traffic-Police-Smartglasses-Read-License-Plates.

Cadwalladr, Carol. (2017, February 26). Robert Mercer: The big data billionaire waging war on mainstream media. *Guardian*. Retrieved online March 22, 2017 at: www.theguardian.com/politics/2017/feb/26/robert-mercer-breitbart-war-on-media-steve-bannon-donald-trump-nigel-farage.

Chamayou, Gregoire. (2015). *A theory of the drone*. New York, NY: New Press.

Chammah, Maurice. (2016, February 3). Policing the future. *The Verge*. Retrieved from www.theverge.com/2016/2/3/10895804/st-louis-police-hunchlab-predictive-policing-marshall-project.

Charlton, James Martin, & Moar, Magnus. (2017). A desire for immersion: The panorama to the Oculus Rift, unpublished manuscript. Retrieved online March 22, 2017 at: http://eprints.mdx.ac.uk/18924/.

Coldewey, Devin. (2017, April 5). Taser rebrands as Axon and offers free body cameras to any police department. *Tech Crunch*. Retrieved online March 22, 2017 at: https://techcrunch.com/2017/04/05/taser-rebrands-as-axon-and-offers-free-body-cameras-to-any-police-department/.

Daley, Jason, Piore, Adam, Lerner, Preston, & Elizabeth Svoboda. (2011, November 14). How to fix our most vexing problems, From Mosquitoes to potholes to missing corpses. *Discovery Magazine*. Retrieved online March 22, 2017 at: http://discovermagazine.com/2011/oct/21-how-to-fix-problems-mosquitoes-potholes-corpses.

Deleuze, Gilles. (1992). "Postscript on the societies of control." *October* 59, 3–7.

Dewey, C. (2016, June 14). Here are Mark Zuckerberg's full remarks about how much he'd like to (literally!) read your thoughts. *Washington Post*. Retrieved online March 20, 2017 at: www.washingtonpost.com/news/the-intersect/wp/2016/06/14/here-are-mark-zuckerbergs-full-remarks-about-how-much-hed-like-to-literally-read-your-thoughts/?utm_term=.f04dc469ed5b.

Foucault, Michel. (2012). *Discipline & punish: The birth of the prison*. London: Vintage.

Garling, Caleb. (2015, November). Virtual reality, empathy and the next journalism. *Wired Magazine*. Retrieved online March 20, 2017 at: www.wired.com/brand-lab/2015/11/nonny-de-la-pena-virtual-reality-empathy-and-the-next-journalism/.

Gitlin, Todd. (1980). The whole world is watching: Mass media in the making & unmaking of the new left. Univ of California Press.

Greenberg, Jonathan. (2015, February 15). The real problem with Harf's "Jobs for Jihadis" program, the *Observer*. Retrieved online March 22, 2017 at: http://observer.com/2015/02/the-real-problem-with-harfs-jobs-for-jihadis-program/.

Jouvenal, Justin. (2016) The new way police are surveilling you: Calculating your threat "score." *Washington Post*, available at www.washingtonpost.com/local/public-safety/the-new-way-police-are-surveilling-you-calculating-your-threat-score/2016/01/10/e42bccac-8e15-11e5-baf4-bdf37355da0c_story.html.

Koepke, Logan. (2016, November 21). Predictive policing isn't about the future. *Slate*. Retrieved online March 22, 2017 at: www.slate.com/articles/technology/future_tense/2016/11/predictive_policing_is_too_dependent_on_historical_data.html.

Latour, Bruno. (2004). Why has critique run out of steam? From matters of fact to matters of concern. *Critical Inquiry*, *30*(2), 225–248.

Ng, Alfred. (2017, April 5). Police hear a pitch for free body cameras, with a side of AI. *Cnet*,. Retrieved online March 22, 2017 at: www.cnet.com/news/police-free-body-cameras-artificial-intelligence-taser-axon-vievu/.

Nguyen, Nicole. (2015). Chokepoint: Regulating US student mobility through biometrics. *Political Geography*, *46*(2015), 1–10.

Paglen, Trevor. (2014, November) Operational images. *E-flux Journal*, 59,. Retrieved online March 20, 2017 at: www.e-flux.com/journal/59/61130/operational-images/.

Rapping, Elayne. (2004). Aliens, nomads, mad dogs, and road warriors: The changing face of criminal violence on TV. In S. Murray, & L. Ouellette (Eds.), *Reality TV: Remaking television culture* (pp. 214–230). New York: New York University Press.

Rowinski, Dan. (2013, November 14). Connected air: Smart dust is the future of the quantified world. *Readwrite*. Retrieved online March 20, 2017 at: http://readwrite.com/2013/11/14/what-is-smartdust-what-is-smartdust-used-for/.

Sledge, M. (2013, March 3). CIA's gus hunt on big data: We "try to collect everything and hang on to it forever." *Huffington Post*. Retrieved from www.huffingtonpost.com/2013/03/20/cia-gus-hunt-big- data_n_2917842.html.

Virilio, Paul. (1989). *War and cinema: The logistics of perception*. London: Verso.

Wall, Tyler. (2013). Unmanning the police manhunt: Vertical security as pacification. *Socialist Studies/Études Socialistes*, *9*(2).

Zizek, Slavoj. (1996). *The indivisible remainder: An essay on Schelling and related matters*. London: Verso.

Žižek, Slavoj (1999). The matrix, or, malebranche in Hollywood. *Philosophy Today*, 43. Supplement (1999),11–26.

6 Algorithmic patrol

The futures of predictive policing

Dean Wilson

'Predictive Policing' has rapidly emerged as one of the new catchphrases of police practitioners. In essence, the broad claim of predictive policing is that through the algorithmic processing of data sets – both crime- and non-crime related – patterns of *probable* future offending and victimization can be revealed and subsequently interdicted prior to their actualization. The mathematical formulas engaged in predictive policing draw upon a confluence of algorithmically driven predictive modelling innovations, emerging from commercial, military, natural disaster and public health sectors, generally relying upon advanced information processing and machine learning. Although the origins of predictive policing extend back to early experiments in the 1970s with computer-assisted crime control, recently it appears to have captured the imagination of police practitioners, elements of the media and the wider public on an unprecedented scale. In 2011 *Time* magazine hailed predictive policing as one of the 50 best inventions of the year (*Time*, 28 November 2011). A 2013 report estimated that over 150 US police departments were using predictive policing software (Bond-Graham & Watson 2013), with the number internationally also indicating notable technological diffusion globally (McCulloch & Wilson 2016: 84–86). With predictive software packages designed for law enforcement materializing with astounding rapidity – and a multitude of police departments globally investing in the technology – increasing recourse to predictive analytics in police work appears highly likely.

The seductive appeal of predictive policing is explicable not only with reference to technological innovation, but additionally in relation to a wider security context increasingly oriented towards the pre-emption of threats prior to their materialization. This orientation became noticeably strident in the wake of the 9/11 attacks, where confidence was placed in the capability of science and technology to guide data-driven pre-emptive interventions that could foil future terrorist incidents prior to their actualization (Lyon 2003). The downward drift of pre-emptive logics into quotidian criminal justice has attracted considerable scholarly attention of late. Zedner (2007) has postulated a shift from Ulrich Beck's conception of the 'risk society' towards a 'pre-crime society', while Amoore (2013) adumbrates a transmutation from notions of the probable towards those of the possible. In this predictive policing represents a hybrid form

– increasingly seeking to target suspect identities and locales prior to any offending, while also mobilizing strategies that emerge from the origins of modern policing such as preventive patrol, and others closely associated with ideas of risk management such as broken windows policing. While predictive policing rallies the discursive tropes of pre-emption, it is also an extension (rather than reinvention) of police practices that have a longer historical trajectory.

The advent of predictive policing is also frequently discursively fused with the rise of 'Big Data'. 'Big Data' refers to more than simply size. It refers to 'size, storage medium and analytic capacity' (Andrejevic & Gates 2014: 186; see also boyd and Crawford 2012). The element of analytic capacity is particularly important in terms of pre-emption, as the application of 'predictive analytics', with the power to mine huge quantities of data, and assemble patterns and correlations previously inconceivable, are increasingly engaged to predict everything 'from the weather to the behaviour of financial markets' (Andrejevic & Gates 2014: 186). While data-mining is both descriptive and predictive, the capacity of such analytics for prediction is increasingly promoted and sought. Both data-mining and predictive analytics have their origins in commerce rather than national security or domestic criminal justice (Gandy 2006). Nevertheless, as numerous authors have noted (Andrejevic 2013; McCulloch & Wilson 2016), there has been considerable seepage – not to mention cooperation – between the agencies of state security and domestic policing and commercial organizations who maintain substantial quantities of 'Big Data' and are versed in its analysis. While there is a clear valence towards Big Data in predictive policing, it should also be acknowledged that some predictive policing software relies on more traditional police data. Big Data and predictive policing are not yet intrinsically synonymous.

This chapter examines the recent history and implementation of predictive policing. Beginning by situating predictive policing within the broader context of information and communication technologies in policing, along with discussion of its fundamental premises and the little that is currently known of its efficacy in reducing crime rates. The discussion then considers the theoretical (and atheoretical) contexts of predictive policing, looking both at claims that predictive policing is beyond theory, and at more critical perspectives applicable to the technology. It will then consider predictive policing as a commodity and market, examining how predictive policing has been astutely marketed as a commodity that betokens institutional professionalism and sophistication. As predictive policing is still comparatively novel, the chapter will conclude by considering some possible consequences for policing and society of predictive policing.

What is 'predictive policing'?

Predictive policing may be defined as 'the application of analytical techniques – particularly quantitative techniques – to identify likely targets for police intervention and prevent crime or solve past crimes by making statistical predictions'

(Perry, McInnes, Price, Smith, & Hollywood 2013: xiii). This has kindled intense speculation that what once seemed feasible only in the imaginings of science fiction - the prevention of crimes *before* they occur - is now within the reach of police organizations. Some observers have questioned whether predictive policing is actually anything new. One of the attendees (a Police Chief from Nebraska) at the National Institute of Justice (NIJ) Predictive Policing Symposium in 2009, for example, suggested that predictive policing was not new, but rather 'a coalescing of interrelated policing strategies and tactics that were already around, like intelligence-led policing and problem solving. This just brings them under the one umbrella of predictive policing' (Pearsall 2010: 8). Nevertheless, the same commentator also concluded that predictive policing's novelty resided in 'the tremendous infusion of data' (Pearsall 2010: 18).

While the events of 9/11 heralded the intensification of the surveillance capacity and informatization of policing, such developments have built upon evident historical trends. There is a significant lineage of technological innovation in policing that has refashioned police organizations and interactions with the public, apparent since the early twentieth century in the adoption of cars and motor cycles, two-way radio, telephones and computer–aided dispatch (CAD) systems (Manning 1992). By the late 1980s computers were intrinsic to the infrastructure of police organizations, and research evidence emerged from the 1990s onwards suggesting that information and communication technologies could *potentially* have significant impacts upon police organization and practice (Ericson & Haggerty 1997; Chan 2001; Chan, Brereton, Legosz, & Doran 2001; Manning 2001, 2003, 2008). While it was often hoped that information and communication technologies would lead to improved efficiencies, ethnographic studies suggested that this remained highly contingent upon local conditions and organizational structures. The variety of technology, the emphasis of individual police units, and the skills and competencies available within police departments mediated the overall shaping of individual socio-technical systems (Chan 2003: 658). Some ethnographies suggested that the actual impact of information and communication technologies upon police work was negligible. Manning's ethnographic research into crime mapping and analysis in three US police forces suggested that the 'penetration of technology into the contours of the job is almost entirely dependent on its perceived utility on the ground', going on to conclude that 'IT and its supporting features did not change any significant practice' (2008: 251). Conversely Ericson and Haggerty (1997) and Chan et al. (2001) suggested more profound – though contradictory and ambiguous – impacts upon policing, including the levelling of hierarchies, information overload, the over-collection of data, and resistance to and subversion of technology by police in the field. While it is too early to assess whether similar trajectories will accompany predictive policing, previous research examining information and communication technologies suggests useful vectors of inquiry.

Predictive policing may be comparable with older communications technologies, such as two-radio communications and telephones, in terms of organizational impact. However a more direct lineage can be traced through the introduction of crime mapping to policing. Although there were initial experiments with crime mapping using Geographic Information System (GIS) software in the 1970s, these were generally only engaged by large (and well-resourced) police agencies, primarily due to the expense of the equipment required and the technical expertise of the personnel necessary to maintain it. Client server technology in the 1980s increased the diffusion of such mapping systems, but only began to proceed apace with the development of personal computers with sufficient space and capacity to handle the software requirements necessary for crime mapping.

These technical precursors of predictive policing were supplemented by the theoretical work of George Kelling, who published a series of papers for the conservative Manhattan Institute in the 1980s addressing the topic of how best to deploy police resources through statistical analysis. In the mid-1990s, such ideas were realized with the introduction of CompStat in the New York Police Department (NYPD) – intended to reorganize policing in accordance with fluctuations in local area crime statistics. While CompStat presaged predictive policing through a close reliance and reaction to statistical data, it is also important as a precursor in terms of the policing strategies mobilized on the basis of that data. CompStat was amalgamated with the 'broken windows theory' of policing, famously associated with Wilson and Kelling (1982), which saw NYPD officers focusing upon minor 'quality of life' violations in the belief that this was would short-circuit more serious violations of the law (Henry 2009).

Nevertheless, while such strategies were mostly in the vein of 'hot spot' identification and analysis, by the late 1990s there were already suggestions that 'ideally, early indicators of troubled areas would help to identify future patterns of crimes committed across time and space and inform a more proactive approach to policing' (La Vigne & Groff 2001: 217). The NIJ subsequently initiated a predictive modelling research programme (sponsoring five projects) using statistical techniques ranging from artificial neural network mapping through to spatial econometrics. Several of these projects explicitly drew upon the 'broken windows' theory of crime control, monitoring minor offending in order to provide statistical evidence of its assumed connection to subsequent and more serious offending (La Vigne & Goff 2001). Despite the appearance of predictive techniques in the late 1990s and early 2000s, initial reception was sceptical, even amongst ardent advocates of crime mapping and crime science. Chainey and Ratcliffe (2005), for example, argued that it was a field still 'very much in its infancy' (178). Even so numerous papers had already appeared that were attempting to operationalize techniques drawn from epidemiology and artificial neural network analysis to what was at that point still predominantly termed crime 'forecasting' (La Vigne & Goff 2002).

The coining of the term 'predictive policing', in tandem with the key principles undergirding it, are generally credited to Police Chief William J. Bratton and the Los Angeles Police Department (LAPD). By 2008 Bratton was prominent in media and policy circles advocating the success of the use of predictive analytics to forecast gang violence and to monitor offending in real time. Wider popularization of the term 'predictive policing' emerged in 2009, largely due to a three-day predictive policing symposium held jointly by the NIJ, the Bureau of Justice Assistance and the Los Angeles Police Department. A second practitioner symposium was held in 2010 in Providence, Rhode Island addressed challenges, successes and limitations, while also pointing to the urgent need for data sharing and robust analytical capacity within police organizations (Perry et al. 2013: 4–5). Notwithstanding more restrained commentators favouring the term 'forecasting' (seen as 'objective, scientific and reproducible', whereas prediction is defined as 'subjective, mostly intuitive and non-reproducible') (Perry et al. 2013, xiii), the term 'predictive policing' has emerged as the accepted terminology for this new generation of data-driven policing technologies.

Critiques of Big Data and algorithmic governance have often pointed to the opacity of predictive calculations as threatening accountability and ideals of transparency. Social scientists are increasingly drawing attention to the socially contingent – and consequently political – construction of algorithms and digital data sets. Aradau and Blake have recently argued that while we can safely assert that there is no such thing as 'raw data'; 'critical data studies need to analyse the 'making up' of data and the production of kinds of data at the juncture between information and computer science, on the one hand, and politics, on the other' (2015: 9). As Introna and Wood indicate, within complex socio-technical networks, technologies such as algorithms 'function as political "locations" where values and interests are negotiated and ultimately "inscribed" in the very materiality of the things themselves – thereby rendering these values and interests more or less permanent' (2004: 180). Nevertheless, with many of the algorithms deployed for predictive policing remaining proprietary, interrogation of the data inputs and how they are processed remains elusive. Predictive policing is therefore increasingly a form of 'black-box' policing. As Latour noted in explanation of the concept of 'blackboxing':

> When a machine runs efficiently, when a matter of fact is settled, one need only focus on inputs and outputs and not on its internal complexity. Thus, paradoxically, the more science and technology succeed the more opaque and obscure they become.
>
> (1999: 304)

In the case of predictive policing, such 'blackboxing' is often cultivated through the design impulse to render the technology 'user-friendly'. Earlier studies of policing and technology have noted the contingency of technological innovations in police organizations (Manning 2001), with heralds of initial promise dissipating within police bureaucracies lacking technical expertise, while also defending traditional notions of policing as a craft. A response to this absence of expertise within policing organizations has been aspire to ever greater usability and simplicity. Writing in 2002, Groff and La Vigne suggested that crime-forecasting might be derailed through complexity and the requirement for specialized technical knowledge, recommending that its diffusion within police organizations would be best facilitated 'through the automation of forecasting techniques into a user-friendly software program' (50). Some vendors have embraced this challenge. Promotional material for Wynyard Analytics – marketing predictive software for medium- and small-scale policing agencies – suggests one of the key advantages of its product is that the 'technical capacity is hidden from the user and the science is already built in' (Wynyard Group: 2). PredPol also promotes its system's 'intuitive usability' suggesting actionable predictions are available with 'one click of a mouse' (www.predpol.com/technology/). As Jonathan Crary has recently argued, the sense that technological developments are somehow quasi-autonomous 'allows many aspects of contemporary social reality to be accepted as necessary, unalterable circumstances, akin to facts of nature' (2013: 36).

Predictive policing incorporates a wide range of methodologies and products with the commonality that all claim that algorithmic processing of data will more accurately predict crime than previous methods. The origins of the algorithms engaged include supply-chain predictive analytics drawn from major retailers such as Walmart (the story of their supply-chain during an impending hurricane has become a powerful meme in predictive policing circles), earthquake prediction, epidemiology and US Department of Defence technological innovations designed to predict the location of Improvised Explosive Devices (IEDs). There are a wide range of analytical tools engaged including single and dual kernel density estimation, regression methods, spatio-temporal analysis and risk terrain analysis (Perry et al. 2013). Predictive policing, in a relatively short time period, has also experienced a frenetic pace of innovation. Not only have the number of predictive policing software vendors expanded globally, but the reach and range of data sources incorporated within systems has also. Earlier systems predominantly focused upon the places of potential property offences – particularly vehicle theft and robbery. More recently systems have become more ambitious incorporating gun crime, drug offences, gang violence, homicide and traffic offences. Increasingly also, as will be discussed later, predictive policing analytics are algorithmically likely offenders and as well likely offences.

There is at present scant independent evaluation evidence that predictive policing reduces crime rates. Nevertheless, media stories are frequently accompanied by statistics suggesting dramatic impacts resulting from the technology. The PredPol website, for example, suggests significant decreases in offending in locations where their software has been engaged; including a 32 per cent drop in burglaries and a 20 per cent drop in vehicle crime in Alhambra California, an 18 per cent reduction in residential burglary in Modesto California, and an aggregate drop in crime of 19 per cent in Atlanta Georgia (www.predpol.com/results/). A recent randomized control trial of the PredPol system also claimed positive outcomes (in terms of declining crime statistics) concluding that their 'model successfully predicted 4.7% of crimes … a predictive accuracy 2.2 times greater than existing techniques' (Mohler, Short, Malinowski, Johnson, Tita, Bertozzi, & Brantingham 2015). By contrast, however, a recent Rand Corporation evaluation of predictive policing in Shreveport, Louisiana found 'no statistical evidence that special operations to target property crime informed by predictive maps resulted in greater crime reductions than special operations informed by conventional crime maps' (Hunt, Saunders, & Hollywood, 2014: xvi). While it is certainly feasible that predictive techniques in policing may have a statistical impact on crime rates, this is by no means established. Critics of the 'broken windows' policing strategy and its mobilization in the 1990s in New York City pointed out that there may well have been other factors presaging declines in crime – including departmental budgets, the priorities of local areas and a range of environmental, social and political factors (Bowling 1999; Karmen 2006;). Similar factors may well be operational in relation to predictive policing. Moreover, there is no evidence that any evaluation has yet attempted to assess the impact of predictive policing upon police–community relations. This would be a worthwhile undertaking, for as Braga, Papachristos, & Hureau note in relation to hot-spot policing 'short-term crime gains … could undermine the long-term stability of specific neighbourhoods through the increased involvement of mostly low-income minority men in the criminal justice system' (2014: 659).

Selling prediction

Commodified technological solutions have proven to be powerfully alluring in contemporary societies (Loader 1999; Zedner 2009; Hayes 2012). For police agencies, predictive policing as a commodity carries with it the aura of high technology, which agencies often view as a 'force multiplier' (Nogala 1995). Moreover, acquiring predictive policing software can enhance organizational prestige, projecting an image of institutional efficiency and professionalism. While PredPol remains the most high-profile of vendors, the sector has witnessed dramatic growth with large technology corporations such as IBM, Palantir, Microsoft and Hitachi all developing predictive software for law enforcement. Moreover there are a host of new start-ups emerging (sometimes subsidiaries of larger enterprises) who have entered the market. Wynyard Group, a company based in Auckland, New Zealand, has made significant inroads with its user-friendly 'Advanced Crime Analytics', which extends the promise of predictive policing capacity to small and medium-sized agencies previously lacking the requisite IT expertise (Wynyard Group 2015: 2). Combining historical police data with social media analysis, their product purportedly allows law enforcement to 'connect the dots and solve the crime – and prevent future crimes' (Wynyard Group 2015: 1).

Predictive policing software has been accompanied by strident marketing that has garnered substantial media coverage. Striking headlines, such as 'Chicago goes Minority Report' (Ernst 2014), advance the conception that the technological capacity to predict future crime is technically feasible – or is at least immanent. Such reports frequently reference Steven Spielberg's 2002 film version of the classic Phillip K. Dick story *Minority Report* (e.g. Vlahos 2012) – a cinematic representation deeply infused in popular culture and mobilized as a signifier for the technological capacity of prediction. Focusing specifically on the marketing for the most well-known system, PredPol, Bond-Graham and Winston

(2013) reported proliferating media interest in predictive policing, much of it recycling 'quotes and statistics drawn directly from press releases written by PredPol for police departments'. Moreover, their analysis of PredPol contracts revealed police departments frequently received discounts if they were prepared to participate in marketing, and were often obliged by the contracts to provide testimonials, appear in the media for case studies and introduce the company to other law enforcement agencies (Bond-Graham & Winston 2013). As their mordantly entitled article concluded 'the future of policing looks a lot like good branding'.

Not only does predictive policing promise to stop crime before it happens, it has also been promoted as a technological means of 'doing more with less'. This provides some explanation for the proliferation of vendors and products appearing under the rubric of predictive policing, particularly during a global economic crisis where a neoliberal doctrine of austerity has been ascendant and its accompanying ideological predisposition for reduced spending on public services (including police departments) has reached unprecedented levels. Consequently, proponents of policing driven through predictive analytics and Big Data have frequently couched its advantages not merely in terms of accuracy, but also in terms of efficiency. Beck and McCue (2009) argue that one of the great strengths of predictive policing is that it 'supports the ability to do more with less, without compromising public safety', in hard economic times.

Despite a paucity of evaluation evidence, policing agencies appear adamant that predictive policing is the material transformation of *Minority Report*–style pre-crime from science fiction to science fact. This outlook was captured in a recent definition offered by Her Majesty's Inspectorate of Constabulary (HMIC) in the UK, who defined predictive policing as 'methods used by police forces to pre-empt crime and prevent it from happening' (HMIC 2014: 73). Similar perceptions of predictive policing are evident in the US, with a recent Rand Corporation report producing a composite letter, indicative of those regularly sent by local police to the International Association of Crime Analysts, that read:

> Dear Sir/Madam: Please let my chief and I know where we can buy the software that will tell us where to go pick up criminals as they are committing crimes. We have read articles and seen ads on this.
>
> (Perry et al. 2013: 127)

Colleen McCue, a key advocate of predictive analytics in law enforcement, while conceding that 'there are no crystal balls in law enforcement and intelligence analysis', echoes comparable faith in predictive policing, suggesting that data mining and predictive analytics 'can help characterize suspicious or unusual behaviour so that we can make accurate and reliable predictions regarding future behaviour and actions' (2005: 57). The enthusiasm with which predictive policing has been greeted by police managers rests to some extent on its promise as a silver bullet for the complexities and problems befalling contemporary policing. Importantly, also, it can be attributed to astute marketing and the general commodification of security.

Policing beyond theory?

In a now commonly cited essay Chris Anderson (former Editor-In-Chief of *Wired* magazine) (2008) enunciated with alarming polemical clarity his clarion call for an epistemological reconceptualization based upon the mammoth processing power and torrents of data emanating from the information economy:

> This is a world where massive amounts of data and applied mathematics replace every other tool that might be brought to bear. Out with every theory of human behaviour, from linguistics to sociology. Forget taxonomy, ontology and psychology. Who knows why people do what they do? The point is they do it, and we can track and measure it with unprecedented fidelity. With enough data, the numbers speak for themselves.

The era of Big Data and predictive analytics is consequently represented as an entirely novel post-theoretical configuration of society, technology and knowledge. According to Mayer-Schönberger and Cukier 'society will need to overturn some of its obsession with causality in exchange for simple correlations: not know *why* but only *what*' (2013: 7). This privileging of correlation over causation, and the attendant instrumentalization of knowledge, has elicited comment from numerous social science scholars, with contemplation and concern evident across disciplines as to the fate of theory and analysis in a Big Data era (boyd & Crawford 2012; Kitchin 2014; Chan & Bennett Moses 2015). Nevertheless, as Chan and Bennett Moses (2015) suggest, there really is no possibility of 'atheoretical' data or analysis, as all interpretations and analysis will be driven by some form of theory, however implicit.

Far from the 'end of theory', Anderson's (2008) exaggerated entreaty for unquestioning acceptance of quantification and computerization, and the privileging of correlation over causality, powerfully evokes a form of instrumental rationality akin to that outlined by the critical theorists of the Frankfurt School. For Max Horkheimer and Theodore Adorno, it was the instrumental rationality resting at the heart of scientific knowledge that resulted in a post-philosophical world, fixated with means over ends. Instrumental rationality, in their view, produced 'neither concepts nor images, nor the joy of understanding, but method, exploitation of the labour of others, capital' going on to conclude that 'on their way toward modern science human beings have discarded meaning. The concept is replaced by the formula, the cause by rules and probability' (2002: 2–3). Similar arguments were advanced by another member of the Frankfurt School, Herbert Marcuse, who argued that 'scientific-technical rationality and manipulation are welded together in new forms of social control' going on to suggest that 'the quantification of nature, which led to its explication in terms of mathematical structures, consequently, separated the true from the good, science from ethics' (1964: 146).

For some critics predictive policing reinvigorates older ideas of police professionalism which epitomized Weberian concepts of bureaucratic rationality, and also gesture towards the instrumental rationality adumbrated by the Frankfurt School. Skalsnky (2011), for example, suggests that the key tenets of classic police professionalism – crime control focus, privileging of objective, scientific decision making and the centralization and rationalization of authority, are returning to the fore, displacing the brief flirtation with community policing of previous decades. From another perspective, the datafication of suspicion inherent in predictive policing resonates with the shift from disciplinary societies to societies of control (Deleuze 1992; Rose 2000), where subjectivities are increasingly replaced by 'data doubles' or 'dividuals' and where rhizomatic networks enact diffused and modulated control through innumerable switch points. More recently Rouvroy (2011) has advanced the concept of 'algorithmic governmentality', networks of increasingly automated control with little interest in prior notions of 'soul surgery' (Foucault 1977). Ericson and Haggerty's formulation of the 'surveillant assemblage' is also pertinent to predictive policing. The quest for data to fuel predictive calculations forges complex rhizomatic networks between public police agencies, other public agencies and the private sector that are often fluid and mutable, and which have complex implications for accountability and the exercise of police power. The contrast between Anderson's claim of knowledge beyond theory, and the critiques of new social configurations of data and automation, reflect the long standing dichotomy between utopian and dystopian visions of technology and society.

Antinomies of prediction

Previous research into the deployment of 'technologies of social control' (Marx 2007) has noted a range of consequences: both positive and negative, intended and unintended. It has already been mentioned that the efficacy of predictive policing in reducing crime is not yet established. However, even if crime reductions were irrefutably proven, this may well only apply to specific locations. Moreover, it is necessary to assess other consequences that might accrue from algorithmically driven police patrol. These include the impacts upon policing practice and autonomy at street-level, the policing strategies engaged by departments in tandem with predictive analytics, the impetus towards augmented police collection of data, and the potential of predictive policing to technologically amplify extant and well-documented discriminatory police practices.

It has been a historical truism that innovations in communication and information technologies in policing have not only increased police potential to exert social control over populations, but have also reflected inwards, increasing bureaucratic control over front-line police (Ericson & Haggerty 1997). Given the current fragility of police legitimacy, particularly in the United States, there is consequently a shift towards engaging data to render policing increasingly visible to the public – a trend already evident through citizens use of social media to document police misconduct (Goldsmith 2010). In the US, one of the principle aims of the Federal *Task Force on 21st Century Policing* is to 'emphasize the opportunity for departments to better use data and technology to build community trust', with the suggestion that data on the use of force, pedestrian and vehicle stops, and police shootings should be made available to the public (Smith & Austin 2015). Predictive algorithms also have the potential to migrate into police management via recruitment and accountability. While many police departments already boast early-warning systems to predict police misconduct, algorithms have recently been developed for this purpose, although an initial experiment with algorithmically predicting misconduct in Chicago was abandoned in the face of resistance from the police union, who pilloried the system as a 'crystal-ball thing' (Arthur 2016). It is also likely that predictive policing technologies will increasingly incorporate the monitoring of police activity, rendering predictive patrol as a perpetual time and motion study of police activity. This potential is indicated by Mohler et al., who advocate engaging in-car GPS 'to provide a greater level of precision and also provide information on officer activity when not on a predictive policing mission' (2015: 1410).

In addition to enhancing bureaucratic control over police, extant research on police use of information technologies suggests it may also have significant impacts on police work, effectively undermining conceptions of policing 'craft' and experiential knowledge of where and when offending is likely to occur through processes of deskilling. Just as the widespread use of GPS has largely eroded capacities for cognitive mapping (Carr 2015), the concept of tacit police knowledge may be erased as police increasingly rely on algorithmically automated visualizations and calculations of suspicious locales and persons. Moreover, increasing reliance upon predictive analytics and data mining to drive police patrol may well result in greater distancing of police from the communities they are policing. David Sklansky suggests that this reinvention of police professionalism can have just such effects, noting that 'a fixation on technology can distract attention from the harder and more important parts of this process, the parts that rely on imagination and judgement' going on to argue that it also deflects attention from 'other critical parts of the contemporary policing agenda: building trust and legitimacy, ensuring democratic accountability, and addressing the enduringly corrosive connections between criminal justice and racial inequality' (Sklansky 2011: 9–10).

Nevertheless, it is sometimes claimed that predictive policing can simultaneously augment – rather than replace – extant police skills, enhance accountability and improve police–community relations. The outcomes of predictive policing for communities policed will be contingent upon the strategies deployed. PredPol suggests that officers patrolling targeted locations will have 'an opportunity to interact with residents, aiding in relationship building and strengthening community ties' (www.predpol.com/how-predpol-works/). In the Los Angeles Police Department Pacific Division predictive policing has been accompanied by the release of predictive maps to the public via social media, who are entreated to assist the police in crime prevention with the message 'You can simply walk with a neighbor, exercise, or walk your dog in these areas and your presence alone can assist in deterring would be criminals from committing crime in your neighborhood' (Friedersdorf 2014). Such responsibilization of individuals into the web of crime control is consistent with the individualized ethos of neoliberalism. It may, furthermore, produce insidious consequences, portraying the city as a dangerous 'fearscape' peopled by threatening immanent predators (Kindynis 2014; see also Wallace 2009) and thereby justifying punitive interventions and legitimating yet further investment in predictive technology.

While predictive policing may potentially involve aspects of community policing, there is little evidence for this at present. Predictive policing appears to inevitably gravitate towards focused patrol in an effort to interdict a range of street-level offences against property and persons. In the same manner as CompStat was aligned with the zero tolerance model of policing (Harcourt 2001; Punch 2007), early commentators on crime 'forecasting' envisaged it as most compatible with a range of criminological approaches that askew wider social explanations; such as routine activities theory and situational crime prevention (Groff & La Vigne 2002: 32). This elective affinity between predictive techniques and asocial models of criminality echoes Garland's observations of the 'criminologies of everyday life' which, he argued, 'offer an approach to social order that is, for the most part, amoral and technological' (2001: 183). The evaluation of predictive policing in Shreveport, Louisiana suggests that it is these 'criminologies of everyday life' that are most commonly drawn upon. The report noted that in one of the trial districts, 'there was a large emphasis on intelligence gathering through leveraging low-level offenders and offences' and that police 'stopped individuals who were committing ordinance violations or otherwise acting suspiciously' (Hunt et al. 2014: 12). As Ratcliffe (2014) suggests, there is little evidence predictive policing has moved beyond directed patrol.

Predictive policing also energizes a fetishization of intelligence. In the 1990s, Ericson and Haggerty had already noted the tendency of information technology to cultivate the 'police over-production of "just-in-case" knowledge about crime and criminals' (1997: 438). If crime is simply a natural phenomenon amenable to prediction and interdiction through data mining, statistical predictions are also envisaged as gathering enhanced precision through the accumulation of ever larger data sets. This tendency is further reinforced by the promoted capacity of data mining and predictive analytics to 'identify unusual or subtle patterns in very large datasets' that exceed 'the analytic capacity of the human brain or even traditional computer-based methodologies' (McCue & Parker 2003). The conviction that inconceivable and counter-intuitive correlations will materialize energizes data collection by police agencies – often data that bears no immediately discernible relationship to crime. Police have been advised 'to tap into the wealth of non-traditional data available locally, such as medical and code-compliance data' (Pearsall 2010: 19). Moreover, the potential for algorithms to discern all manner of previously inconceivable correlations leads to the relentless pursuit of perpetually expanding data sets. As one article on predictive policing notes, 'everything from the timing of gun shows to the weather and the phase of the moon is deemed potentially important' (Vlahos 2012). Nevertheless it is not clear that sheer volume will produce any tangible benefits. Police data generally remains fragmented, with disparate information systems and error-prone data sets. Augmenting these with vast flows of private, semi-public and public data drawn from disparate agencies and commercial enterprises may well simply exacerbate errors and inaccuracies, simultaneously making such flaws more difficult to detect.

Such incessant data compilation threatens to render crime statistics – which are notoriously problematic – even more so, as opaque data streams form statistical feedback loops that assume their own concretization masked behind the algorithmic calculations of predictive policing. As Pasquale (2015) notes, it is quite difficult to find a *less* objective set of statistics than crime figures. Put simply 'high crime' areas attract more police, who perform more arrests and collect more data from an area, which consequently feeds back into the statistical profile which then fabricates an even higher crime area – and so on. As Pasquale goes on to note:

> Once that set of 'objective' data justifies even more intense scrutiny of the 'high crime' neighbourhoods, that will probably lead to more arrests – perhaps because of a real crime problem, but perhaps instead due to arrest quotas or escalating adversarialism between law enforcement and community members.
>
> (2015: 42)

Possibly, therefore, predictive policing will simply technologically amplify extant discriminatory practices. This seems plausible in relation to race and policing – an area where there is ample research demonstrating discriminatory practices (Brunson & Miller 2006; Alpert, Dunham, & Smith 2007). Despite the claim that an order fabricated through Big Data, predictive analytics and machine learning is purely technical and neutral – unable to manifest racism or prejudice as the algorithm is incapable of base human prejudice – this seems extremely unlikely to unfold in practice. Rather it seems apparent that extant patterns of discriminatory policing will be rendered less visible beneath the high-technology sheen of machine calculation. Additionally, as Harcourt (2007) argues, such patterns of stigmatization label whole communities and geographic locations as suspect, and energize troubling spirals that amplify suspicion.

It also notable that predictive policing increasingly focuses not only on aggregates but also upon individual identities deemed liable to immanent transgression. The Chicago Police Department received a US$2 million grant from the National Institute of Justice in 2009 to develop predictive policing. They subsequently compiled a heat list of the '400 most dangerous people' at risk of violence, to be visited by police *before* any criminal act is committed. By 2016 the list, officially dubbed the 'strategic subject list' and based on algorithmic correlation of 11 variables and allocating a risk score from 1 to 500 – had expanded to 1400 names (Rhee 2016). Police, members of the community and social service agencies have utilized home visits (termed 'custom notifications) to visit 'at risk' subjects prior to any offence. The list, according to the developer

Miles Wernick, was compiled 'in an unbiased, quantitative way' (Stroud 2014). Nevertheless the 'strategic subject list' is concentrated in areas with high Latino and Black populations, who are consequently the subjects of intensive surveillance. The Shreveport experiment also engaged individual targeting through parole records and juvenile arrest records (Hunt et al. 2014).

Predictive policing, through its proclamation of calculative precision which increasingly drills down to identify specific individuals, extends the allure of what Valverde and Mopas (2004) term 'targeted governence' – a security utopia whereby finely calibrated policing decisions are enacted with ever greater precision. Additionally predictive policing is instrumental in the formation of the rhizomatic 'surveillant assemblages' outlined by Haggerty and Ericson (2000), with police agencies functioning as 'centres of calculation' for disparate flows of data and actions that span and connect public and private agencies. Predictive policing elevates the task of 'crook catching' as the primary aim of police work. It additionally does so in a context where policing itself is becoming increasingly militarized (Balko 2013). Moreover, while there is much reference to the natural sciences in terms of the origins of predictive policing, it is significant that many of the tools currently developed were initially for military application, and that the founders of PredPol had previously undertaken significant predictive projects for the military (Bond-Graham & Winston 2013). This opens up an even more disturbing vista through which to interpret predictive policing – one that is strikingly vivid if we look at the opportunities for tracking and targeting that potentially unfold with the advent of 'real time' predictive analytics fused with mobile patrol. Kade Crawford of the American Civil Liberties Union of Massachusetts recently decried the trend towards 'para-militarizing our police, turning them all into robo-cops who take their directions from computers as to how to go about their day' (cited in González 2015). As McCulloch and Wilson note, the tools of predictive policing:

> rather than generating novel and unthinkable predictions, risks projecting the all-too-thinkable racialized and discriminatory policing practices of the present into the future, with the inevitable result that they energize a spiralling data-driven loop of racialized, militarized and punitive pre-emptive interventions.
>
> (2016: 85)

While some predictive policing products exclusively utilize crime statistics compiled by police, there is increasing interest in social media data to facilitate predictive capacity and propelling predictive policing towards intensified engagement with Big Data. Hitachi's Visualization Predictive Crime Analytics, for example, promotes the system's capacity to use 'natural language processing for topic intensity modelling using social media networks together with other public and private data feeds in real time to deliver highly accurate crime predictions' (Hitachi 2015). Experiments conducted at the University of Virginia on social network analysis extol the virtue of analysing tweets, with researchers reporting that tweet mining and analysis could predict 19 out of 25 crime types, and furthermore going on to suggest that where tweets contained no direct information about crime, they 'may contain information about activities associated with them' (Gerber 2014; Lever 2014). Funded by the US military in the hope that such technology would enable the prediction of threats in the battle zones of Iraq and Afghanistan, such developments underscore the technological continuum that ranges from commercial to military applications. One leading producer of 'smart' weapons, multinational security contractor Raytheon, has also developed an 'extreme-scale analytics' system RIOT (Rapid Information Overlay Technology), which mines social media networks. A Raytheon representative has suggested that RIOT facilitates the translation of dispersed social data into 'useable information' that will 'help meet our nation's rapidly changing security needs' (Gallagher 2013).

Other technological innovations taken up within policing agencies have facilitated fresh capacities for data collection and surveillance. One of the more controversial of these in the US has been the deployment of International Mobile Subscriber Identity (IMSI) catchers to capture identifying information and individual locations from mobile phones. The commercial brand 'Stingray' has found most favour with police agencies, and is reportedly in wide use across the US, although its legality in terms of search and seizure has been called into question. The issue with such technologies is that they collect information not only on suspects, but also provide police with access to the mobile phones of everyone in a particular location. Some versions of the Stingray also boast the capacity to not only collect mobile phone IDs, but also to access numbers dialled on individual mobiles and even phone conversations (Harwood and Stanley 2016; see also Monahan 2016: 236–237).

The increasing integration of open-source intelligence (OSINT) and particularly social media intelligence (SOCMINT) is particularly challenging for police agencies, but is nevertheless evermore drawn upon as a source for producing predictions of crime and disorder. Social media does, however, represent complex problems of representativeness and accuracy, and while there have recently been efforts to account for bias in social media analysis – such as the model of 'tension monitoring' developed by Williams, Edwards, Housley, Burnap, & Rana, et al. (2013) few police departments have adequate expertise to undertake analysis internally. This often leads to reliance on external private providers to process data and supply information, leading to obfuscation of the origins of data and a blindness to its flaws (Trottier 2015). The tendency to incrementally expand data sets in relation to predictive policing has already been noted. With SOCMINT such temptations are multiplied, giving rise to surveillance creep and the collection and analysis of vast swathes of information – potentially traceable to individuals. The mobilization of SOCMINT in the policing of protest and dissent underscores the potential democratic hazards. This has been evident in the UK where the vague term 'domestic extremism' has been engaged to justify the monitoring of a wide range of groups from anti-austerity protestors and animal rights activists. The National Domestic Extremism

Unit, with a staff of 17 working 24 hours a day, was reported to be scanning tweets, YouTube videos, Facebook and other social media sources, engaging tools from the private sector such 'sentiment analysis' and 'horizon scanning' (Wright 2013; Denick, Hintz, Carey, & Pandya 2015). How far such instances of 'high' policing will percolate into routine 'low' policing is uncertain. However faith that more data enhances predictive precision is likely to propel extensive data collection that can be stored and algorithmically interrogated for future patterns.

The futures of prediction

Predictive policing may not inherently involve a drift towards authoritarianism and 'broken windows' orientated law-enforcement. A more benign aspect of predictive policing is evident in efforts to engage predictive analytics to locate incidents of hate crime (Williams et al. 2013), and there are additional potentials for algorithmic police accountability that are only just being explored. Additionally there is a danger that critical scholars may become as entranced as police managers with the marketing rhetoric of predictive policing, imbuing the technology with transformational capacities that rarely materialize in practice. Limitations in staffing and organizational acceptance of the technology on policing's front line may well mediate some of its more deleterious impacts – and some potentially positive ones also. There are already instances where predictive policing has been abandoned altogether. In Milpitas, California predictive policing was abandoned after three years, as, the Police Chief explained in an interview 'we often did not have sufficient staff to post officers at PredPol identified locations and still remain responsive to priority calls for service' (Bauer 2016). It is likely that the evolution of predictive policing will be highly contingent upon the specific organizational cultures where it is engaged.

Nevertheless, predictive policing has troubling aspects that frequently run counter to notions of democratic accountability and social justice. The ceaseless thirst of predictive policing to incorporate data fragments from diverse public and private sources, combined with the propriety nature of many algorithms, veers towards a dangerous opacity and multiplies the chances of vulnerabilities and errors; and the possibility of these very flaws in data being obfuscated within black-boxed systems. Moreover the frequent hyperbole, promotion and appeals to neutrality and the objectivity of machine learning and predictive analytics elides the prejudicial, and frequently irrelevant, nature of the fragments fed into the multitude of security algorithms. The technological sheen of predictive analytics also conceals the capacity of predictive policing to digitally reinscribe and amplify racialized and militarized policing tactics, shielding entrenched patterns of racial profiling behind the new mode of algorithmic calculation. And yet the most profound and problematic consequences of predictive policing could reside in the transformation of the very nature of policing. Predictive policing – despite references to its potential to enhance police–community relations and trust – more often presents an instrumental vision of policing as pure crime control.

Unfortunately, all too often this dovetails with the view police have of themselves as crime fighters rather than social workers. Moreover it potentially amplifies the most negative aspects of contemporary policing such as racialized and militarised policing, while eclipsing 'social service' tasks and public communication that could increase trust in communities. Far from a neutral technical tool, predictive policing's present trajectory presages intensified punitive policing increasingly distanced from those policed.

References

Alpert, G., Dunham, R., & Smith, M. (2007). Investigating racial profiling by the Miami-Dade Police Department: A multimethod approach. *Criminology and Public Policy*, *6*, 25–56.

Anderson, C. (2008, 23 June). The end of theory: The data deluge makes the scientific method obsolete. *Wired Magazine*. www.wired.com/science/discoveries/magazine/16-07/pb_theory.

Andrejevic, M. (2013). *Infoglut: How too much information is changing the way we think and know*. London: Routledge.

Andrejevic, M., & Gates, K. (2014). Big data surveillance: Introduction. *Surveillance & Society*, *12*(2), 185–196.

Aradau, C., & Blanke, T. (2015). The (big) data-security assemblage: Knowledge and critique. *Big Data & Society*, *2*, 1–12.

Arthur, R. (2016, 9 March) *We now have algorithms to predict police misconduct: Will police departments use them?* http://fivethirtyeight.com/features/we-now-have-algorithms-to-predict-police-misconduct/.

Balko, R. (2013). *Rise of the warrior cop: The militarization of America's police forces.* New York: Public Affairs.

Baranuik, C. (2015, 14 March). Caught before the act. *New Scientist*.

Bauer, I. (2016, 11 June) Police: Tech contract meant to predict, prevent crime in Milpitas nixed. *Mercury News*, www.mercurynews.com/milpitas/ci_30115970/police-tech-contract-meant-predict-prevent-crime-milpitas.

Beck, C., & McCue, C. (2009). Predictive policing: What can we learn from Wal-Mart and Amazon about fighting crime in a recession? *The Police Chief*, 76, www.policechiefmagazine.org.

Bond-Graham, D., & Watson, A. (2013, 30 October). All tomorrow's crimes: The future of policing looks a lot like good branding, *SF Weekly*.

boyd, d., & Crawford, K. (2012). Critical questions for big data: Provocations for a cultural, technological and scholarly phenomenon. *Information, Communication & Society*, *15*(5), 662–679.

Bowling, B. (1999).The rise and fall of New York murder: Zero tolerance or crack's decline? *British Journal of Criminology*, 39, 531–554.

Braga, A., Papachristos, A., & Hureau, D. (2014). The effects of hot spots policing on crime: An updated systematic review and meta-analysis. *Justice Quarterly*, *31*(4). 633–663.

Brunson, R., & Miller, J. (2006). Gender, race and urban policing: The experience of African American youths. *Gender and Society*, *20*(4), 531–552.

Carr, N. (2015). *The glass cage: Who needs humans anyway?* London: Vintage.

Chainey, S., & Ratcliffe, J. (2005). *GIS and crime mapping.* Chichester: Wiley.

Chan, J. (2001). The technological game: How information technology is transforming police practice. *Criminal Justice*, *1*(2), 139–159.

Chan, J. (2003). Police and new technologies. In T. Newburn (Ed.), *The handbook of policing* (pp. 655–679). Cullompton: Willan.

Chan, J., Brereton, D., Legosz, M., & Doran, S. (2001). *E-Policing: The impact of information technologies on police practices.* Brisbane: Queensland Criminal Justice Commission.

Crary, J. (2013). *24/7: Late capitalism and the ends of sleep.* London: Verso.

Delueze, G. (1992). Postscript on the societies of control. *October*, 59, 3–7.

Denick, L., Hintz, A., Carey, Z., & Pandya, H. (2015). *Managing 'threats': Uses of social media for policing domestic extremism and disorder in the UK.* Cardiff: Cardiff School of Journalism, Media and Cultural Studies: Cardiff.

Ericson, R., & Haggerty, K. (1997). *Policing the risk society.* Toronto: University of Toronto Press.

Ernst, D. (2014, 20 February). Chicago goes 'Minority report': CPD big on predictive policing. *Washington Times*, www.washingtontimes.com.

Foucault, M. (1977). *Discipline and punish: The birth of the prison.* London: Penguin.

Friedersdorf, C. (2014, 28 March). To prevent crime, walk the dog on at-risk blocks. *The Atlantic.*

Gallagher, R. (2013, 10 February). Software that tracks people on social media created by defence firm. *Guardian.*

Gandy, O. Jr. (2006). Data mining, Surveillance and discrimination in the post 9/11 environment. In K. Haggerty, & R. Ericson (Eds.), *The new politics of surveillance and visibility* (pp. 363–384). Toronto: University of Toronto Press.

Garland, D. (2001). *The cultural of control: Crime and social order in contemporary society.* Oxford: Oxford University Press.

Gerber, M. (2014). Predicting crime using Twitter and Kernel Density Estimation. *Decision Support Systems*, *61*, 115–125.

Goldsmith, A. (2010). Policing's new visibility. *British Journal of Criminology*, *50*(5), 914–934.

González, R. (2015). Seeing into hearts and minds: Part 2: 'Big data', algorithms and computational counterinsurgency. *Anthropology Today*, *31*(4), 13–18.

Griffith, D. (2015, 30 June). Predictive policing: Seeing the future. *Police*, www.policemag.com/channel/technology/articles/2015/06/predictive-policing-seeing-the-future.asp.

Groff, E., & La Vigne, N. (2002). Forecasting the future of predictive crime mapping. *Crime Prevention Studies*, *13*, 29–57.

Haggerty, K., & Ericson, R. (2000). The surveillant assemblage. *British Journal of Sociology*, *51*(4), 605–622.

Harcourt, B. (2001). *Illusion of order: The false promises of broken windows policing.* Cambridge, MA: Harvard University Press.

Harcourt, B. (2007). *Against prediction: Profiling, policing and punishing in an actuarial age.* Chicago: University of Chicago Press.

Harwood, M., & Stanley, J. (2016, 19 May). *American military technology has come home-to your local police force.* www.thenation.com/article/American-military-technology-has-come-home-to-your-local-police-force/.

Hayes, B. 2012. The surveillance-industrial complex. In K. Ball, K. Haggerty, & D. Lyon, (Eds.), *Routledge handbook of surveillance studies* (pp. 167–190). London: Routledge.

Henry, V. (2009). Compstat. In A. Wakefield, & J. Fleming (Eds.), *The Sage dictionary of policing* (pp. 49–52). London: Routledge.

Hitachi (2015). Hitachi data systems unveils new advancements in predictive policing to support safer, smarter societies. Press release, 28 September, www.hds.com/en-us/news-insights/press-releases/2015/gl150928.html.

HMIC (Her Majesty's Inspectorate of Constabulary). (2014). *Policing in austerity: Rising to the challenge compendium*. London: HMIC.

Horkheimer, M., & Adorno, T. (2002) *Dialetic of enlightenment: Philosophical fragments*. Stanford, CA: Stanford University Press.

Hunt, P., Saunders, J., & Hollywood, J. (2014). *Evaluation of the Shreveport Predictive Policing Experiment.* Santa Monica, CA: Rand Corporation.

Introna, L., & Wood, D. (2004). Picturing algorithmic surveillance: The politics of facial recognition systems. *Surveillance & Society*, *2*(2/3), 177–198.

Karmen, A. (2006). *New York murder mystery: The true story behind the crime crash of the 1990s.* New York: New York University Press.

Kindynis, T. (2014). Ripping up the map: Criminology and cartography reconsidered. *British Journal of Criminology*, *50*(2), 222–243.

Kitchin, R. (2014). Big data, new epistemologies and paradigms. *Big Data & Society*, *1*, 1–12.

Latour, B. (1999), *Pandora's hope: Essays in the reality of science studies*. Harvard University Press.

La Vigne, N., & Groff, E. (2001). The evolution of crime mapping in the United States: From the descriptive to the analytic. In A. Hirschfield, & K. Bowers (Eds.), *Mapping and analysing crime data: Lessons from research and practice* (pp. 203–221). London: Taylor & Francis.

Lever, R. (2014, 21 April). Researchers use Twitter to predict crime. *Sydney Morning Herald*, www.smh.com.au.

Loader, I. 1999. Consumer culture and the commodification of policing and security. *Sociology*, *33*(2), 373–392.

Lyon, D. (2003). *Surveillance after September 11.* London: Polity.

McCue, C. (2005). Data mining and predictive analytics: Battlespace awareness for the war on terrorism. *Defence Intelligence Journal*, *13*(1&2), 47–63.

McCue, C., & Parker, A. (2003). Connecting the dots: Data mining and predictive analytics in law enforcement and intelligence analysis. *Police Chief*, *70*(10), 115–122.

McCulloch, J., & Wilson, D. (2016). *Pre-crime: Pre-emption, precaution and the future.* London: Routledge.

Manning, P. (1992). Information technologies and the police. In M. Tonry, & N. Morris (Eds.), *Modern policing* (pp. 349–398). Chicago: University of Chicago Press.

Manning, P. (2001). Technology's ways: Information technology, crime analysis and the rationalization of policing. *Criminal Justice*, *1*(1), 83–103.

Manning, P. (2003) *Policing contingencies.* Chicago, Ill: University of Chicago Press.

Manning, P. (2008) *Crime mapping, information technology and the rationality of crime control.* New York, NY: New York University Press.

Marcuse, H. (1964). *One-dimensional man: Studies in the ideology of advanced industrial society.* Boston, MA: Beacon Press.

Marx, G. (2007). Rocky bottoms: Techno-fallacies of an age of information, *International Political Sociology*, *1*(1), 83–110.

Mayer-Schönberger, V., & Cukier, K. (2013). *Big data: A revolution that will transform how we live, work and think.* London: John Murray.

Mohler, G., Short, M., Malinowski, S., Johnson, M., Tita, G., Bertozzi, A., & Brantingham, P. (2015). Randomized controlled field trials of predictive policing. *Journal of the American Statistical Association*, *110*(512), 1399–1411.

Monahan, T. (2016). Built to lie: Investigating technologies of deception, surveillance and control, *The Information Society*, *32*(4), 229–240.

Nogala, D. (1995). The future role of technology in policing. In J.-P. Brodeur (Ed.), *Comparisons in Policing: An International Perspective* (pp. 191–210). Aldershot: Ashgate.

Pasquale, F. (2015). *The black box society: The secret algorithms that control money and information.* Cambridge, MA: Harvard University Press.

Pearsall, B. (2010). Predictive policing: The future of law enforcement? *NIJ Journal*, *266*, 16–19.

Perry, W., McInnes, B., Price, C., Smith, S., & Hollywood, J. (2013). *Predictive policing: The role of crime forecasting in law enforcement.* Santa Monica, CA: RAND Corporation.

Punch, M. (2007). *Zero tolerance policing.* Bristol: Policy Press.

Ratcliffe, J. (2014). What is the future … of predictive policing? *Translational Criminology.* Spring, 4–5.

Rhee, N. (2016, 2 June). Can police big data stop Chicago's spike in crime. *Christian Science Monitor*, www.csmonitor.com/USA/Justice/2016/0602/Can-police-big-data-stop-Chicago-s-spike-in-crime.

Rose, N. (2000). Government and control. *British Journal of Criminology*, *40*, 321–339.

Rouvroy, A. (2011). Technology, virtuality and utopia: governmentality in an age of autonomic computing. In M. Hildebrant, & A. Rouvroy (Eds.), *Law, human agency and autonomic computing: The philosophy of law meets the philosophy of technology* (pp. 119–140). London: Routledge.

Skalsnky, D. (2011). The persistent pull of police professionalism. *New Perspectives in Policing*. Washington DC: Harvard Kennedy School/National Institute of Justice.

Smith, M., & Austin, R. (2015). Launching the police data initiative. *The White House President Barack Obama*, 18 May, Retrieved from: https://obamawhitehouse.archives.gov/blog/2015/05/18/launching-police-data-initiative.

Stroud, M. (2014, 19 February). The minority report: Chicago's new police computer predicts crimes, but is it racist? *The Verge*. Retrieved from: www.theverge.com.

Trottier, D. (2015). Open source intelligence, social media and law enforcement: Visions, constraints and critique. *European Journal of Cultural Studies*, *18*(4), 530–547.

Valverde, M., & Mopas, M. (2004). Insecurity and the dream of targeted governance. In W. Larner, & W. Walters (Eds.), *Global governmentality: Governing international spaces* (pp. 232–250). London: Routledge.

Vlahos, J. (2012). The department of pre-crime. *Scientific American*, *306*(1), 62–67.

Wallace, A. (2009). Mapping city crime and the new aesthetic of danger. *Journal of Visual Culture*, *8*(1), 5–24.

Williams, M., Edwards, A., Housley, W., Burnap, P., Rana, O., Avis, N., Morgan, J., & Sloan, L. (2013). Policing cyber-neighbourhoods: Tension monitoring and social media networks. *Policing & Society*, *23*(4), 461–481.

Wilson, J., & Kelling, G. (1982, March). Broken windows: The police and neighbourhood safety. *Atlantic Monthly.*

Wright, P. (2013, 23 June). *Meet prism's little brother: Socmint.* www.wired.co.uk/socmint.

Wynyard Group. (2015). *Wynyard advanced crime analytics: Powerful software to prevent and solve crime*. Auckland: Wynyard Group.

Zedner, L. (2007). Pre-crime and post-criminology? *Theoretical Criminology*, *11*(2), 261–281.

Zedner, L. (2009). *Security.* London: Routledge.

Part IV

Automated justice

7 Algorithmic crime control

Aleš Završnik

> Big data analytics have the potential to eclipse longstanding civil rights protections in how personal information is used in housing, credit, employment, health, education, and the marketplace.
>
> (The White House, 2014)

> The discriminatory and even predatory way in which algorithms are being used in everything from our school system to the criminal justice system is really a silent financial crisis. (Speaking of the usage of Big Data today and the predatory lending practices of the subprime crisis in 2008).
>
> (Cathy O'Neil, 2016)

Introduction

Big data promises to solve the problems of society and individuals, ranging from cybersecurity and police investigations, to enhancing the quality of judicial decision-making, product customisation and, personalisation, from marketing strategies and targeted advertising, to self-monitoring and lifestyle improvement. The EU Commission, for instance, anticipates the benefits of big data in domains as varied as health, food security, climate change, resource efficiency, energy, intelligent transport systems, and smart cities (European Commission, Towards a thriving data-driven economy, COM(2014) 442 final, Brussels, 2 July 2014).

The big data industry claims that algorithmic decision-making eliminates biases inherent in human decisions. Judges may not be aware of biases and mental shortcuts ("heuristics") in their decision-making processes. They may have only little insight into the disparities in their judgements due to facts as trivial as the influence of nutrition on their decisions (Danziger, Levav, & Avnaim-Pesso, 2011). On the other hand, the industry claims algorithms are impartial in the process of applying the abstract rules in a particular case. Turkle's study (Turkle, 1995) of people's attitudes towards hypothetical computer judges show how people may even prefer computer judges when the judicial system is regarded as being unfair.

The promises of big data, predictive algorithmic calculations, and the related Artificial Intelligence (AI) in policing and in the process of delivering justice surely seem, at least at first sight, impressive. For introductory purposes, let us turn to merely two examples. Researchers claim, for instance, that AI can draw inferences about a person more accurately than one's friends and colleagues (Youyou, Kosinski, & Stillwell, 2015). When computer-based judgments were focused only on the Big Five Personality Test traits (i.e. openness, conscientiousness, agreeableness, extraversion, and neuroticism), these correlated more strongly with participants' self-ratings than average human judgments do (Youyou et al., 2015). Such computer programs can also be very exact in hypothesising about the inclinations of a particular judge, and psychological big data research may in this way enter into courtrooms in order to assess witness testimony etc. (cf. application Apply Magic Sauce, a personalisation engine that predicts psychological traits from digital footprints of human behaviour).

In policing, PredPol, one of the leading crime predictive software companies, claims to predict crime in advance. The algorithm was developed with more than a decade of police crime data and over a period of six years. The program shows coloured "hotspots" for locations where and when crime is likely to occur on the basis of data on previously reported crimes. In a 21-month experiment, co-authored by a co-founder of the PredPol company (Mohler, Short, Malinowski, Johnson, Tita, et al., 2015), the algorithm used data on burglaries and thefts from and of cars from Los Angeles and data from Kent, UK, where patterns for those crimes, as well as violent crimes including assault and robbery, were analysed. The researchers tested the computer model by pitting it against professional crime analysts. The results show that ETAS models (near real-time epidemic-type aftershock sequence crime forecasting) predict 1.4–2.2 times as much crime compared with a dedicated crime analyst using existing criminal intelligence and hotspot mapping practice.

Similarly as predictions in other contexts, such as predictive analytics for product recommendations made by Amazon, movies by Netflix, or music by Spotify, AI has been entering the courtroom, and the offices of legal counsellors and the parole board. AI has been used to predict the outcomes of legal problems (Ashley & Brüninghaus, 2006), to decide on co-financing litigations (Legalist.us), and to help human lawyers conduct research more quickly (rossintelligence.com). They have been tested as regards predicting the judicial decisions of the highest courts: the Supreme Court of the USA (Kravets, 2014) and the European Court of Human Rights (ECtHR) (Aletras, Tsarapatsanis, Preoţiuc-Pietro, & Lampos, 2016).

What this chapter claims is that there is intuitive value-based knowledge that should not be overruled by algorithms regardless of the volume of data and the veracity of the data crunching. Crime is a social construct and responses to crime inherently carry values, sentiments, and perceptions as regards what ought to be. These can never be fully grasped by an even continuously evolving "self-learning" algorithm. Algorithm building, e.g. data set preparation and the cleansing of data, on the one hand, and the interpretation of the results thereof, on the other, are carried out by humans and for humans. These two steps cannot escape human error, bias, values, interests, and a human framing of the world regardless how well intended they may be. Big data needs big human control (cf. Buttarelli, 2015). Big data needs big protection (cf. (Marr, 2016). The deployment of big data "solutions" have to be subjected to transparent procedures (Pasquale, 2015) as these "solutions" can never represent life in all its complexity (Morozov, 2013).

The chapter begins with the fascination with numbers, which started at the "birth" of criminology as a science in the nineteenth century: How and why have understanding and tackling crime – an inherently normative phenomena – always relied on non-normative "value-free" numbers? The chapter then presents contemporary critiques of "traditional" nation-oriented crime surveys, and continues by showing how big data attaches to penal power. Then it focuses on automated policing programs and automatisation of other criminal justice actors by presenting several programs that are already being used or tested. The negative effects of automated policing and automated criminal justice are analysed in the concluding section.

Crime control and the fascination with numbers

In the nineteenth century, the emerging modern state with its strong regulatory ambitions found, in statistics, the knowledge to legitimatise the process of disciplining the "dangerous classes" (Garland, 1985). Distinct criminological "schools" addressed the promise of understanding crime differently, but the formation of the central question – what is crime and how to understand crime in order to prevent it – stands as the cornerstone of the foundation of criminology in the nineteenth century, when it leaned heavily on the rising psychiatric movement and then emerging new type of knowledge – statistics (Garland, 1985: 112) It was statistics that was regarded as the scientific knowledge necessary to understand crime, while other types of knowledge penetrated into the criminal justice system only later in the mid and late twentieth century. For instance, in figuring out the motives of the serial killer Norman Bates, one of the detectives in Hitchcock's *Psycho* claimed: "If anybody gets any answers, it'll be the fellow talking to him now … the Psychiatrist."[1] At the beginning of the twenty-first century, the "ultimate" knowledge that criminal justice actors should rely on changed again and instead of psychoanalytically informed psychiatrists, neuroscientists and geneticists stepped in. While genetics and neuroscience still focused on the human being, with a shift in focus from the psyche to the body, today, with the advent of algorithms, aggregates and profiles extracted from populations have become the mode. "The question of crime" got attached to algorithmic techniques, "[which] draw together possible relations and associations across data items from otherwise unrelated databases" (Amoore, 2014: 428) in order to form actionable "data doubles".

Crime statistics carried the promise of attaching reason to modern criminal justice systems and detaching therefrom the "irrational" cruelty of the *ancien régime*. Crime can be measured, the legitimacy of police and criminal justice actors can be studied, and also predicting recidivism and rehabilitation success can be forecasted. This initial fascination with numbers can be traced back to two criminological schools.

First, it was Quetelet who applied statistics to social science in the late nineteenth century and called it "social physics". His goal was to understand the statistical laws underlying social phenomena (Beirne, 1987). In *Essay on Moral Statistics in France* (1832), Quetelet, together with Guerry, relied on statistical techniques to gain insight into the relationships between crime, on the one hand, and poverty, education, and alcohol consumption, on the other. They invented the first "data visualisation" tools, the so-called choropleth maps, which showed how these variables varied across France. These tools were limited to human perception itself as the granularity of such maps was limited by the ability of the human eye to distinguish five to seven colour categories, but despite the deficiencies, the method of flattening-down crime to two dimensional charts and tables enabled the modern state to grapple with crime.

As suggested by Foucault (1995), the new science of criminality helped fortify the penal power of the modern state. Positivist criminology emerged as a very calculated response to the needs of penal power. It was a discourse designed to justify new penal strategies, i.e. training the soul through "moral rehabilitation" with the help of the deprivation of liberty. Criminology served to legitimise and extend modern penal power by presenting punishment as a scientific procedure (Garland, 1992). It could only serve such purpose by leaning heavily on statistics.

Second, the other stream of thought relevant for understanding the contemporary penetration of big data into modern penal power comes from insights pursued by the Chicago School (i.e. the Ecological School), which specialised in urban sociology. Contemporary computerised systems such as COMPAS (Correctional Offender Management Profiling for Alternative Sanctions, a probation assessment instrument) are grounded in a process that started at the beginning of the twentieth century, when the social ecologist Burgess started calculating the recidivism rates of parolees. In 1927, the Chicago School's urban sociologists calculated recidivism rates based on data on 3,000 parolees from Illinois (Harcourt, 2015). Recidivism probabilities have been deduced from "objective" parameters and presented with eye-catching visualisation tools ever since.

The two streams of thought demonstrate how responding to crime has always been connected with a specific type of knowledge that was regarded – at a particular time and space – as culturally "respectable" and perceived as "scientific". The project of building a modern state was deeply dependant on "objective" indicators that statistics was able to offer. The aspiration to depict society in an objective fashion enabled progressive ideas to be joined with statistics: "Large, complex issues could now be surveyed simply by scanning the data laid out geometrically across a single page" (Davies, 2017).

While this fascination with numbers in the nineteenth century is part of the disciplinary power of the modernisation process (Foucault, 1978; 1990), today's fascination with algorithmic calculations in the big data age is part of the neoliberal turn as regards controlling the masses and the changing roles and position of nation states in governing the world. Scholars have depicted this process by claiming, for instance, that we are not living in a disciplinary society, but in a "society of control" (Deleuze, 1992), with a schizophrenic type of surveillance that stems from above ("panopticism"), from below ("synopticism") (Mathiesen, 1997), and/or that even develops like "weeds", thus forming a "surveillant assemblage" (Haggerty & Ericson, 2000). The changing roles and position of nation states have significantly changed, they are sharing their powers with digital corporations. Authors claim that a "military-internet complex" (Harris, 2015), wherein government agencies are joining with tech giants, such as Google and Facebook, to collect vast amounts of information, as well as a "surveillance-industrial complex" (Ball & Snider, 2013), have come to the fore. This digital turn is making modern statistics compiled by nationally funded central agencies obsolete. The data Facebook gets from its unpaid workers, i.e. its "users", far supersede the volume, depth, and immediacy of the data collected by national statistical bureaus. This "new oil" is more detailed and granulated, it offers insight into thoughts and emotions, the "heartbeat" of users, their sentiments and resentments, while the "old type data" focused on the now fading boundaries of state and regions without more granulated details about personal lives (Davies, 2017). For instance, by measuring GPD per capita for the whole country or region, states can only paint a grand picture and say very little about a particular individual or family, while big data juggernauts can paint a granulated and detailed picture of their users. Accordingly to this new balance of power, the governance has thus changed.

The functioning of the state has become dependent on digital juggernauts that collect enormous quantities of data as part of the mundane activities of their users. Thus, "actuarial discipline", as depicted by Feeley and Simon (Feeley & Simon, 1992), is today facing a digital twist and acceleration: vast amounts of all sorts of data produced by smart cars, smart homes, smart grids, wearable computers, etc., are being transformed into real-time actionable data for those who can afford it (Andrejevic, 2013). In the domain of crime and security, instead of psychological and sociological insights into crime, which supported the "Foucauldian" disciplinary power, mathematics took the stage. Mathematical reasoning is offering a new language that is being used for security purposes (Amoore, 2014). Mathematics coupled with computer science is now regarded as a "scientific", "value-free", fault-proof, and "objective" corpus of knowledge that the leading class should rely on. Amoore succinctly claims: "Mathematics promises the calculability of security problems" (Amoore, 2014: 426). In other words, mathematical reasoning and mathematical language is supposedly offering more "comprehensive" and the utmost "actionable" picture about crime.

What we can then observe is how the once "respectable" traditional statistical methods have lost the aura of "objective" and all-encompassing knowledge. Policy makers are now more fascinated with algorithmic power, which is generating new insights into the populace in general, and delinquency in particular, with a rapid increase in ready-to-use "actionable" data (Lane, Stodden, Bender, & Nissenbaum, 2014: xiii). However, with the development of big data techniques, concerns about the "flattening" of the social sphere – inherent in every statistical operation – have not dissipated. New tools are making subjectivity and narrative in policing and judicial legal reasoning obsolete. What counts as "proper", effective, and efficient police work and judicial decision-making has changed. Calculating "things before they happen" on the basis of vast amounts of varied types of data, not just samples of data, which can be operationalised for immediate action is a promise that big data and algorithmic masters are happy to keep. Big data analytics has become the "proper" knowledge of the post-modern, affluent, populist elite.

Post-truth and humanist critiques of crime statistics

Statistics helped take the first step in the flattening of the social world including crime. Observed phenomena were filtered and simplified into indicators and displayed in eye-catching charts and tables. This enabled governments to "circumvent the need to acquire broader detailed local and historical insight" (Davies, 2017). The complexity and fluidity of the social realm was reduced to manageable and comprehensible facts and figures.

The implicit choices as regards what is included in and what is excluded from statistical calculations have always been a political issue inherent in the use of statistics (Desrosières, 2002). While statistics should not be denied its significant contribution to humanistic advances in crime control, statistical reasoning has also revealed its limits in crime control as two types of attacks on statistics and its "expert" knowledge have showed: (1) one from the populist right, resulting in a more emotion-based governance of society that relies heavily on proprietary forms of focused, individualised knowledge derived from big data; and (2) the other from humanist scholars depicting "expert" (statistical) knowledge as reductive and negligent of culture, locality, subjectivity, and narrative.

The latter, humanist critique, claims that statistics is flattening the social realm and reducing crime to uniform categories emptied of substance. But crime is a normative and never a value-free phenomenon. Statistics also neglects individual criminal narratives and the uniqueness of the parties involved in crime. Statistical knowledge is too abstracted from lived experience and thus statistics should not be seen as the ultimate evolution in the understanding of crime, but merely as the least common denominator on which modern states have too heavily depended. This point becomes strikingly obvious if we view the penal system as a regulator of not only crime but also of poverty and social inequality, as so succinctly discussed by Wacquant (Wacquant, 2009). The "evidence-based" policy making may have the aspiration to detach decisions from arbitrary and biased ground, but it has exerted a degree of naïveté. As discussions on the capacity of numbers between Wittgenstein and Turing show, "mathematics [...] is always already political precisely because of its combined faculties of intuition and ingenuity" (Amoore, 2014: 428–429). From an empirical perspective, it is clear the "evidence-based" paradigm has not come close to being able to build a just society – just compare this to the 2017 Oxfam report on wealth disparities, which claims that just eight white men own the same amount of wealth as half the world (Hardoon, 2017).

As an advocate of humanistic education and a critic of "expert" knowledge, Liessmann claims that the neoliberal turn has pushed forward miseducation (German: *Unbildung)*, which lacks the transformative element of education (Liessmann, 2006). This miseducation prioritises "expert knowledge" confined to narrowly defined fields of study. It focuses on the needs of the "free market", the "utility of knowledge", and not the autonomy of the subject, i.e. the sovereignty and maturity of the individual. In such a system, the virtues are flexibility and ease of adaptability to the immediate market requirements. There is no need to see the wider context and there is no room to ask unpleasant questions that go against stereotypes (Liessmann, 2006). In this sense, also statistics is reductive in understanding crime and should be complemented with other types of knowledge focusing on subjectivity, locality, and the micro scale of crime.

This attack on statistical expertise should then be distinguished from the "post-truth" populist critiques and "post-narrative" politics that are utterly diminishing the value of statistical insight. While the simplifications inherent to statistical endeavour have mostly been recognised and other value-driven approaches have found their way into understanding crime, the contemporary "post-truth" paradigm is more radical and promises to displace the need for statistical expertise and even comprehension whatsoever – at least for those with access to the data by using big data and algorithms. But in fact, new knowledge extracted from big data reinforces the new neoliberal powers. The shift from the logic of statistics to that of data is thus accompanied by a shift in governance. In this "post-truth" and "post-narrative" framing of the world, agencies in the security and control domain are expected to become more flexible. For instance, the police are under pressure "to do more with less". It is under pressure to embrace the "gig" economy and Uber-style management, and big data companies are ready to supply such management tools and "solutions".

Big data science and penal power

While in modernity statistics was used in pursuit of the public interest, today, public governance relies on crunching data with the help of data analysts who primarily pursue their own benefits over the public interest. The contemporary rise of populism and reinforcement of the neoliberal "slim state" fit well with the advent of big data analytical tools for the affluent. Digital big data companies accumulate data to meet their ends. The algorithms they develop to make sense of large quantities of data are their central assets – just think of Google and its algorithmic power to offer results from search engine queries, or Uber's power, which is not vested in cars (hardware) but in algorithms (software). The danger of a "post-statistical society" (Davies, 2017) is then not a lack of any forms of truth or expertise altogether, but because these has become privatised and available only in order to empower the ruling minority to achieve its ends. Big data analytics thus – willingly or not – reinforces the neoliberal "post-truth" populist paradigm.

The clandestine operations of proprietary big data analytics are obviously in stark contrast to modern statistical expertise and polling based on public and transparent collection of data by public administrations and the subsequent public availability of such data. These methodologies are not developed for the study of society, in the manner that statistics in modernity was. Predictions are made without any public accountability as big data companies are reluctant to hand over either data or algorithms. The people are informed of big data operations only by accident, sporadically, through data leaks or mistakes that these companies make. There is only little transparency in the procedures that these companies follow in calculating their results. Let us think, for instance, of the power of Facebook, when news-feed manipulation in 2014 influenced its users' sentiments (Kramer, Guillory, & Hancock, 2014), or similarly when ProPublica revealed racial biases in police algorithms in the US only after engaging their own data scientist and without having access to the proprietary algorithm (Angwin, Larson, Mattu, & Kirchner, 2016). The new knowledge-power nexus is changing the governance of society, which also affects policing and the criminal justice system.

Automated policing

Big data has been entering the crime control domain through four actors: (1) intelligence agencies, which have been mandated to provide not only broadly defined national security, but also to curb the most severe types of crime, such as terrorism; (2) law enforcement agencies; (3) criminal courts through remand and parole procedures; and (4) probation commissions in the phase of executing criminal sanctions.

However, big data has changed these actors' daily operations differently. While the police and intelligence communities have been more enthusiastic in accepting preventive tools in their work, the judiciary has been more reluctant to adopt them. Nevertheless, differences exist within police agencies themselves, where police management has been more prone to use such tools in a climate of political pressure "to achieve more with less" and to adapt to shrinking budgets. On the other hand, police officers have been more opposed to the use of such technologies. "Techno-policing" stands in contrast to the traditional role police officers were trained for, such as to patrol neighbourhoods and enter communities and overall being more on the street than in the police station (Leman-Langlois, 2008). The cultural differences between countries have also influenced how sentencing guidelines, IT-based judicial-support systems, and parole prediction programs are perceived. Especially Anglo-American criminal justice experts have been more prone to use such systems, and practitioners have also been politically forced to use guidelines or sentencing information systems in their work (Franko Aas, 2005).

Let me now turn to specific programs in the crime control domain by focusing on what I call "automated policing" and the "automated criminal justice system". The promises of big data analytics as regards national security and law enforcement agencies have been significant: it can be used for:

> identifying threats and crimes before they happen, finding critical information faster, detecting associations between people and activities, improving the accuracy of threat and crime analysis, enabling information sharing and collaboration between investigative organisations, protecting sensitive facilities from attack and preventing emergent cybersecurity risks.
>
> (Akerkar, Vega-Gorgojo, Løvoll, Grumbach, Faravelon et al., 2015)

The definition is appalling and includes (1) police and intelligence work; (2) preventive activities and the investigation of crime already committed; and (3) a narrower focus on certain types of crime (e.g. cybercrime).

Algorithmic "cops" preventing crime

The predictive mapping systems used by the Detroit Police Department shows how police can act proactively. When predictive mapping started showing an increase in auto theft and larceny in the city centre, Detroit's 1st Precinct got together and laid a trap (Smith IV, 2015):

> So the officers set out bait cars wired with cameras, sensors and GPS tracking devices. In the summer, the maps started to turn up bike theft, and the precinct set up bait bikes using the same techniques. In both cases, the criminals were caught. Crime was reduced. The police officers could watch the hotspots fade on the predictive map's interface.

The promise that big data tools would "see things before they happen" and the neoliberal principle of "doing more with less" public funds increased the automation of routine police work such as crowd-monitoring enhanced with plate and face recognition technology and also other domains of police work in several countries. For instance, the American Terrorism Information Awareness Program gathers and analyses data on foreign terrorists in order to pre-empt and defeat terrorist acts, the result of which is more commonly known as the infamous "no fly lists". The American program called TrapWire seeks patterns indicative of a terrorist attack. It completes a prognosis by using data from CCTV installed all across a town, and combines facial recognition features with data from police databases. Other countries are using such tools too: for instance, the Australian border management aims at identifying potential visa regulation violations well before foreigners reach the physical borders of the country. The Dutch police analyse the mobility data of known drug traffickers in order to enhance the chance of arresting them in transit. The London police have been enhancing community policing with a "system optimisation tool" that tracks 30 sources of information, assembles them in groups, identifies citizens' sentiments, and identifies influential individuals. The tool identifies common topics from users' blog posts and 3.3 million tweets per day. If popular discontent dangerously escalates towards social unrest, police can interfere proactively, e.g. by conversing with locals. Similarly, the London Metropolitan Police have tested a criminal gang-scanning program. This predictive crime prevention software was used to identify members of criminal gangs that are very likely to commit violent crimes (Kelion, 2014). The five-month pilot study, which used police data from the previous five years and data from social media posts, enabled more efficient allocation of resources and not only identified dangerous groups but made connections between particular individuals. The police claimed the program did not assign "guilt" by association, but helped predict where gang member violence was likely to occur. IBMs Blue Crush program is an example of police predictive software that has taken the idea of "seeing things before they happen" to its limit. The software produces probability reports on crime yet to be committed by specifying the date and place of future crime. Hitachi's Visualization Predictive Crime Analytics program has been rolled out in half a dozen US cities (as reported in 2015 by Fast Company [Captain, 2015]): "A human just can't handle [it] when you get to the tens or hundreds of variables that could impact crime [...] like weather, social media, proximity to schools, Metro stations, gunshot sensors, 911 calls."

However, police agencies are not the only ones using big data analytics for crime prevention purposes. The application Beware, which seeks to empower first responders with "enhanced situational awareness" (West Corporation, 2016) creates "awareness of potential threats by utilising public information about people, places, and properties that may not be generally available or quickly accessed in local, state, and/or federal criminal databases." The Beware application provides actionable intelligence and assesses risk by "sorting and scoring billions of commercially available records such as vehicle registrations, criminal records, warrants, property records, and known associates before arrival at the scene". It then calculates "threat scores" within seconds of an initial query. The score supposedly indicates a person's potential for violence (Jouvenal, 2016), but examples of misuse have been reported (Burris, 2016). For instance, Fresno Councilman Clinton Olivier, a libertarian-leaning Republican, asked for his name to be run through the system and came up as a "green" which indicates that he is "safe". When they ran his address, however, it popped up as "yellow", meaning the officer should beware and be prepared for a potentially dangerous situation. Perhaps a person that lived in the house before the Councilman was responsible for raising the threat score.

Other actors using big data analytics include securities market regulators and tax administrations. The US Securities and Exchange Commission, for instance, uses data mining to focus more precisely on sectors and companies at risk of engaging in illegal activities. Canadian authorities have been applying big data analytics for tax fraud prevention purposes – so-called "robo taxes" are being designed for the future development of automated taxation systems. Although, *stricto sensu*, tax evasion may not be entirely prevented, losses can be made up for by linking individual consumption with tax data. For instance, the ex-chief of the Financial Administration of the Republic of Slovenia proposed how the administration should link spending patterns with declared personal income tax declarations and use the existent legal possibility to impose 75 per cent taxation on undeclared personal income.

In terms of the volume and speed of data, financial crime is one of the most appropriate types of crime for the application of big data analytics.

> While volume can be the key to recognising fraud, velocity is the key to preventing fraud. [...] New real-time big data platforms enable companies to process massive quantities of historical information and validate new transactions in real-time to spot patterns and halt a transaction before it occurs. By having real-time data at their fingertips, data scientists can also look at new information on the fly, evolving countermeasures just as criminals are adapting to security that's already in place.
>
> (Akerkar et al., 2015: 33).

Algorithmic "cops" solving crime

Big data analytics is being used to investigate crime already committed, for instance in payment fraud and money laundering investigations, which require the analysis of vast amounts of financial transactions data. IBM is offering to banks data analytic tools for combatting skimming card fraud – 15 per cent of all credit card fraud. When financial data is stolen, the bank performs a regression analysis of billions of transactions in order to identify traders who accept non-authorised credit card data and to carry out an analysis of all transactions such merchants have made within a certain period of time. This enables banks to discover connections between scamming and dealer teams and learn about the overall organisation of the scheme.

The gaming industry was one of the first industries to conduct regression analyses of customers, i.e. those who win top payouts. For instance, in the event of a jackpot, the casino carries out a background check in several databases, including those for payment card transactions, and correlates the data with the data in other databases, such as hotel reservation data on the casino's employees, in order to find a "hit". This means it will attempt to ascertain whether the jackpot was actually coincidental, despite the fact, for instance, that the gambler and a casino employee had stayed at the same resort at the same time at some point in the past.

Big data analytics has also changed the investigation of crimes that require the tracking or analysis of prohibited items, tools, and materials. Interpol uses big data analytics to track gun usage across Europe. Investigators of weapons-related crime have access to data related to ammunition and types of weapons, which can facilitate cross-border crime investigations. A computer program called Odyssey analyses data on weapons and types of ammunition (Harris, 2014). Similarly, sexual abuse images and videos of children can be investigated much more efficiently with the use of analytical tools. The NetClean forensic tool, for instance, deals with a vast dataset comprising still images, video, HTML material, and text. Cases involving use of the International Child Sexual Exploitation Image Database (ICSE DB), managed by Interpol, have shown how law enforcement agencies can facilitate the identification of victims and/or perpetrators through the analysis of, for instance, furniture and other mundane items in the background of abusive images or unidentifiable background noise in videos (Plesničar & Klančnik, 2015). Big data capabilities coupled with visualisation tools can detect and compare the similarities of prohibited materials. Similarly, crowdsourcing methods have evolved in the process of tackling child abuse crime. Not only the police but also citizens have been drawn into fighting crime. The TraffickCam app enables travellers to submit pictures of hotel rooms around the world and the information in these images is then matched against a database of images of abused and trafficked children (Huda, 2016). Similarly, CCTV systems enhanced with audio recording capabilities are increasingly dependent on algorithms that can distinguish different noises on the street. By using acoustic fingerprints, early responders can identify the types of weapons used, the place, and the direction of shooting.

The implicit presumptions of automated policing

The above list of manifold uses, actors, and crime-specific tools shows how automated policing was itself developed as a response to the specific cultural milieu that it helped to fashion and in turn reinforce. It is based on several assumptions:

1 The presumption that future crime is calculable, knowable, and targetable before it is committed. The idea is to unfold the future before it happens. As Ulrich Beck (U. Beck, 2002: 40) puts it: "As soon as we speak in terms of 'risk', we are talking about calculating the incalculable, colonising the future." By looking forward "to see things before they happen" this can outperform traditional forms of policing and retrospective crime mapping.
2 The presumption that responding to crime is a sort of management process, where an "in-depth" analysis of the causes of crime is less important than "doing something right here and now". The "fight" against crime ought to mimic the logic of managerialism and focus on how to disrupt the tactics and the "production cycle" of crime.
3 The presumption that algorithmic systems can make judgments about what the data means and that no human analyst can fully grasp and weigh hundreds of thousands of factors in ways only IT can. Algorithms that combine historical and up-to-date crime information are supposed "to do the work of hundreds of traditional crime analysts and produce real-time targeted patrol areas" (Office of the Assistant Attorney General, 2014).
4 The presumption that economic reasoning should be a model for responding to crime and policing in a neoliberal "slim state" should mimic the logic of Silicon Valley start-ups: to use "smart" technology to outsmart the villains,

by increasing the element of surprise and penetrating an adversary's decision cycle and changing outcomes (C. Beck & McCue, 2009).

The urge to develop automated policing (Saunders, Hunt, & Hollywood, 2016) is tightly linked to the "evidence-based" paradigm in policing (Sherman & Weisburd, 1995). It is linked to the neoliberal turn when shrinking police budgets means that the police have to ensure the same level of protection of the public, but with limited personnel, equipment, and training resources. As the Chief of Detectives of the Los Angeles Police Department explained (C. Beck & McCue, 2009).

An automated criminal justice system

Algorithmic pre-trail prediction

Calculating the risks of misconduct has entered into the criminal courts in several ways. Risk assessments are becoming increasingly popular as a way to help set bail for inmates awaiting trial. Several algorithms have been tried out and tested. For instance, the Arnold Foundation algorithm was being rolled out in 21 jurisdictions in the USA as of 2015 (Dewan, 2015). Based on the crunching of data from 1.5 million criminal cases, researchers found that fewer than ten objective factors – basically age, criminal record, and previous failures to appear in court, with more recent offences given greater weight – were the best predictors of a defendant's behaviour. The Arnold Foundation released its methodology, but there was no rigorous audit performed on the program (S. Kramer, 2017).

Another study of pre-trial cases conducted by Stanford University scholars showed that a computer can predict whether a suspect will flee or reoffend – two legal reasons for ordering pre-trail detention – better than a human judge (Leskovec, 2015). Algorithmic prediction was conducted on 1.36 million pre-trail detention cases (Kleinberg, Lakkaraju, Leskovec, Ludwig, and Mullainathan, 2017): (1) data on 360,000 bail cases from the Kentucky federal state, where judges decided to release 73% of the suspects under bail, of which 17% subsequently re-offended or did not return for required court appearances; and (2) federal data on one million criminal bail cases. These two types of "unwanted events" – recidivism or fleeing – were considered in the study as a result of the judges' poor predictions. The researchers tested whether a computer can more accurately make predictions from these 1.36 million pre-trail detention cases and their 40 objective variables, such as the age of a suspect at his or her first arrest, the severity of the crime, the number of past arrest warrants, etc. The goal was to produce prediction models, more specifically, the question was: while Kentucky judges fail to predict the unwanted behaviour of 17% of suspects released on bail, can an algorithm do any better? The research affirmed the hypothesis: the computer achieved a better score and a lower percent of unwanted events at every release level of suspects. In fact, the computer algorithm was on average approximately 20% better at predicting the future behaviour of suspects released on bail. However, while appearing very persuasive, decisions may in fact be less just.

The cases very clearly show the inevitable need for *ad infinitum* improvements. Namely, a human judge may predict the undesired outcomes very well, but would still decide to free someone on bail. For instance, perhaps a judge would deem the fact that a suspect takes care of his/her children to be crucial in deciding not to deprive him/her of liberty, all the while knowing very well that the person might reoffend. There will always be "extra" facts in a particular case that may be unique and go beyond the 40 (or add-any-other-number) parameters but which can crucially determine the outcome of the deliberation process. With self-improving capabilities, algorithms may increase the number of parameters relevant for adjudicating each case, but this can be done *ad infinitum* as two cases are never alike.

Algorithmic sentencing

Actuarial risk assessments have become relatively standard practice in the criminal system, also for correctional placement and in the sentencing phase. Risk assessment in the sentencing procedure was utilised long before the development of ICT tools, but automated tools for determining prison sentences and applying risk and needs assessment (RNA) tools in the sentencing process are relatively newer practices. The central idea of risk assessment in the sentencing procedure is that offenders considered low risk should be given shorter prison sentences than they would otherwise or even avoid incarceration entirely and be granted parole instead, while those deemed high risk should spend more time in prison.

Today, around 60 risk assessment tools are in use across the USA (Barry-Jester, Casselman, and Goldstein, 2015), while the continental legal systems have been lagging behind for at least two reasons. First, the use of calculated risk goes against the existing rules of substantive criminal law that sentences should be determined in relation to two factors oriented towards the past: (1) the degree of criminal liability of the offender at the time of committing an offence, and (2) the severity of the crime committed. Perhaps even more crucially, second, this is due to less governmental interference into judicial independence and autonomy. The dangerousness of an offender or the calculated risk of reoffending may still be taken into account in continental legal systems in the phase for determining the sentence through the use of rules on mitigating and aggravating circumstances.

One of the crucial weaknesses of such prediction tools is the vicious-circle and self-fulfilling prophesies effect. Similar to algorithmic policing, where neighbourhoods identified as risk-prone gain more police attention and thus the police detect more crime there, which in turn leads to the over-policing of such communities, the sentencing stage of criminal procedure captures the same logic. The past behaviour of a certain group decides the fate of an individual, who is, needless to say, a unique human being, with a specific social background, education, skills, a specific degree of guilt, and the specific motives that lead to him/her to commit a crime.

Sentencing algorithms have been tested for biases, and have revealed alarming results. In a detailed assessment of the COMPAS recidivism algorithm, ProPublica (Angwin et al., 2016) looked at more than 10,000 criminal defendants in Broward County, Florida, and compared the predicted recidivism rates with the rate that actually occurred over a two-year period. The COMPAS makes prediction on the basis of a questionnaire containing 137 questions completed by defendants when they are arrested (Maybin, 2016). Their answers are then fed into the software to generate several scores, including predictions of the "risk of recidivism" and the "risk of violent recidivism". Propublica discovered how the system is biased against certain groups: "The formula was particularly likely to falsely flag black defendants as future criminals, wrongly labelling them this way at almost twice the rate as white defendants." (Angwin et al., 2016). In forecasting who would re-offend, the algorithm correctly predicted recidivism for black and white defendants at roughly the same rate (59% for white defendants, and 63% for black defendants), but made mistakes in very different ways. It misclassified the white and black defendants differently when examined over a two-year follow-up period. The case revealed yet another deficiency as regards the interpretation of algorithmic results. The algorithm did not take race directly into account, as the programmers were aware of the basic rules on non-discrimination. The algorithm instead used data that served as a proxy for correlative information: Seemingly neutral language, for instance, have your relatives been sent to jail or prison, target those already living under institutionalized poverty and over-policing. Predominantly, those people are people of colour (Angwin et al., 2016).

Let me now turn to the final stage of the criminal justice system where algorithms have been used intensively to predict criminal behaviour.

Algorithmic probation

Predicting the criminal behaviour of prisoners in order to decide which prisoners should be released on parole can be traced to the pre-WWII period, when parole and probation agencies and commissions used post-conviction assessments to determine the best supervision and treatment strategies for offenders in an effort to reduce the risk of recidivism (Office of the Assistant Attorney General, 2014: 3).

The rehabilitative model of sentencing, which was at the core of the creation of the modern penitentiary, which dominated sentencing and correctional policy until the neoliberal turn in the late twentieth century, was fundamentally based on predicting future behaviour. The goal of the rehabilitative model was to predict when the offender's return to the community would be safe. This was often supported by psychological insight: the gathering of data and considering them through the lens of individual human experience and wisdom in order to arrive at an assessment. It was a human analysis. Psychological research on such human analysis revealed several flaws regarding the "intuition" of human predictors of post-imprisonment misconduct (Office of the Assistant Attorney General, 2014: 3).

Such insights led the U.S. Parole Commission, for instance, to issue guidelines on its release decisions and thus it began to transform parole decision-making to take advantage of statistical modelling. The Commission created the so-called Salient Factor Score, based on such modelling, to help determine what the Commission called "the parole prognosis" (Office of the Assistant Attorney General, 2014: 3).

The sentencing reform movement of the 1980s replaced the rehabilitative model with a new sentencing framework based on truth-in-sentencing and the idea that a criminal sentence should largely be based on the crime committed. The idea that "nothing works" (Martinson, 1974) in 231 studied rehabilitation programs and the idea that there was excessive discretion in charging, sentencing, and parole decisions, led policy to slowly shift towards just deserts. Deterrence and fairness, it was thought, will be accomplished only in such a manner.

The new focus on re-entry has brought with it a renewed need to identify those offenders most at risk of reoffending. This led to the use of the Post-Conviction Risk Assessment (PCRA) Instrument developed by the US federal courts, which is intended to help probation and parole officers determine the level of supervision required for an inmate upon release. The PCRA uses information from an offender's past to identify both the risk of reoffending and the needs to be addressed to lessen that risk. This was a "step in bringing data and the scientific method to corrections" (Office of the Assistant Attorney General, 2014: 5). Other types of predictions were thought to be flawed. More reliable methods not based on subjectivity were regarded as being superior. Once again, criminal justice systems leaned on knowledge that – at a particular time and place – decision-makers perceived as "superior", "more objective", and "more reliable", i.e. actuarial risk assessment and data analytics.

The focus changed from crime to risks. For instance, in Philadelphia (USA) the probation service uses algorithms to predict the recidivism of parolees. The service employs 295 employees who oversee almost 50,000 individuals. The algorithm first calculates which of them are more likely to commit a crime while on parole in the next two years. Each parolee then receives a computer generated "risk score" – low risk, medium risk, or high risk – which is decisive for the type and intensity of oversight during the parole period. Supervisors in charge of supervision are allocated duties accordingly: supervisors of parolees rated as low risk oversee up to 400 individuals, while those who supervise parolees regarded as having a high risk of re-offending oversee only 50 such individuals. Initially, the algorithm was designed to analyse 100,000 cases on the basis of 36 parameters, such as age, sex, and area of residence of the offender, the neighbourhood of *loci delicti commissi*, and previous convictions. But over the years the database was extended and now the algorithm uses crime data from the last 50 years. Specific crimes are analysed for additional predictors and attributes in order to teach the algorithm to predict specific behaviour. The author of the algorithm explains: "I use tens of thousands of cases to build the system, [as well as] asymmetric costs of false positives and false negatives, real tests of forecasting accuracy, the discovery of new forecasting relationships, and yes, machine learning" (Labi, 2012).

Other examples include calculating the risk of persons under community supervision reoffending (Berk, 2008), models assessing the risk that a gang affiliate will be involved in violence as a function of their social relationships (Papachristos, 2007), as gang-related murders occur through an epidemic-like process of social contagion as competing groups jockey for positions of dominance (Green, Horel, and Papachristos, 2017).

However, views differ on the contribution of the Post Conviction Risk Assessment (PCRA) tool in ensuring more just and racially non-segregated results. Today, actuarial instruments have evolved away from race as an explicit predictor. However, this trend was accompanied by two others that would have significant race-related effects: (1) a general reduction in the number of predictive factors used and, (2) an increased focus on prior criminal history (Harcourt, 2015). Criticism has especially focused on the importance and weight given to the criminal history of an offender. It is among the strongest predictors of arrest and is also included in the sentencing guidelines of jurisdictions in the USA, while researchers have shown that criminal history is a proxy for race (Harcourt, 2015). In other words, heavy reliance on criminal history in sentencing contributes more to racial disparities in incarceration than reliance on other robust risk factors less bound to race (Frase, 2009). What should be recognised is that even though the guideline "objective" calculations do not include prohibited attributes, such as race, some of these prohibited attributes are included anyway by "proxy", i.e. criminal history in this case. The "vicious circle" of algorithmic prediction, as Harcourt puts it, proceeds thusly:

> The combination of these two trends – narrowing and focusing on prior criminal records – has proven devastating to African-American communities – and can only continue to have disproportionate impacts in the future. The reason is, the continuously increasing racial disproportionality in the prison population necessarily entails that the narrower prediction instruments, focused as they are on prior criminality, are going to hit hardest the African-American communities.
>
> (Harcourt, 2015: 8)

On the other hand, a study in which the data on more than 34,000 federal offenders were examined to test the predictive validity of the PCRA tool revealed little evidence of test bias in the PCRA: the instrument strongly predicts arrest for both black and white offenders and the resulting score has essentially the same meaning, i.e. the same probability of recidivism, across groups. Instead of speaking of criminal history as a "proxy", the authors claim it is better construed as a "mediator" (Skeem and Lowenkamp, 2016):

"We cannot infer causality from associations, but our results are consistent with what we would expect to see if a causal path leading from race to criminal history to violent future arrest were in force."

The results of this study are in fact modest; the authors merely claim that there is no empirical basis for assuming that the status quo – i.e. judges' intuitive consideration of offenders' likelihood of recidivism – is preferable to judicious application of a "well-validated and unbiased risk assessment instrument" (Skeem and Lowenkamp, 2016). The problem, which the authors do not address, is how to recognise a "well-validated instrument", and why judges should be replaced in the first place.

Conclusion

The transformation of an increasingly greater part of our activities into a digital language creates an enormous quantity of personal data ("big data") upon which the functioning of societies in the age of datafication depends. The predominant opinion is that the thus created big data will influence our way of thinking about the world and our place in it (Mayer-Schönberger and Cukier, 2013). This enthusiasm is shared with "solutionism" (Morozov, 2013), the belief that the difficulties our societies face all have technological solutions, which grows into an ideology, which needs to be addressed especially where the rights and duties of people are involved.

In the "control and security domain", algorithms are replacing human judgment with varying intensity. The principles of big data are used by intelligence agencies, the military, and law enforcement. Drone "signature strikes" manifest how weapons are used on the basis of huge amounts of data, with which algorithms supposedly make "surgically precise" identification of terrorists, i.e. based solely on movement patterns, mobile phone location data, and indicators as mundane as washing lines – as The Freestone Drone, a video installation by George Barber, lucidly showed in 2013: if washing lines are full, the house may more likely be targeted.

In the field of policing, algorithmic policing is extending the relatively static ideas of the crime mapping of "hot spots" (see Weisburd, Bernasco, Bruinsma, 2009) and "actuarial justice" (c.f. the PredPol program and IBM's Blue Crush IBM, 2011). This entails "transcending the general linear reality" (Hope, 2015). In comparison to relatively static "actuarial justice", as described in the famous article by Feely and Simon (1992), such algorithms lead to *automation* in systems of crime control. In the administration of criminal justice, algorithms for deciding on pre-trail detention are being tested (Leskovec, 2015), and algorithms for assisting parole boards are already in use in several US states (Labi, 2012; Skeem & Lowenkamp, 2016).

Advocates of algorithmic courts have been proclaiming how algorithms can make sentencing decisions more just and how many more mitigating and aggravating circumstances can be considered in determining a sentence in a particular case. In the age of big data, reliance on judges' decision-making as to offenders' likelihood of recidivism is presented as less transparent and less consistent than computer-supported, evidence-based risk assessment: the question is which "black-box" is preferred: the one with a human judge or the digital variety?

Algorithmic sentencing is, after all, about the division of power in a democratic society. Predictive policing (Perry, McInnis, Price, Smith, & Hollywood, 2013) is changing according to changes in social and economic conditions. Information sentencing systems have changed power relations by transferring power from the judiciary, as "the least dangerous branch", to distant administrative bodies. Thus, Franko Aas claimed that sentencing has been transformed into "sentencing-at-a-distance", causing more injustice than previously anticipated (Franko Aas, 2005). However, this has been done with the tacit approval of the judicial decision-makers, as they often like to use risk assessment tools not because they are "reliable" predictors, but because using them minimises their own risk of being blamed for the subsequent consequences (Harcourt, 2015). Thus, it concerns the diffusion of accountability among several decision-making bodies, as well as the scientific aura that algorithmic decisions have succeeded in gaining. Similar to "evidence-based policing" (Sherman & Weisburd, 1995), the process of sentencing by algorithm, referred to as "evidence-based sentencing" (Starr, 2013), carries with it weaknesses and threats to fundamental liberties.

The turn towards algorithmic knowledge-power is an attack on subjectivity in the name of supposedly more objective, "scientific", "actionable", reliable, and value-free knowledge, which only digital behemoths can discern. The transition from narrative and database (Franko Aas, 2005) to automated algorithmic justice and algorithmic policing threatens civil liberties and the democratic division of power. The automatisation threatens to eliminate existent human-based narrative and subjectivity altogether and can in fact hinder legal evolution. The most progressive judicial decisions of the highest courts have always been outliers by default. Such decisions radically change the direction of the evolution of legal doctrines. Furthermore, the security decisions based on mathematical calculations are already always political (Amoore, 2014). Additionally, researchers have also pointed out that the "powerful" are still benefiting from personalised services based on human judgement, although this is being cross-checked by means of big data analytics, while algorithms alone are making decisions affecting the lives of the masses (Pasquale, 2015). Such discrimination is erecting new inequalities in the form of an "algorithmic divide". In fact, many are claiming that big data insights can threaten democratic process – by substituting democracy with "algocracy" (Morozov, 2013).

Note

1 *Psycho*, Hitchcock (1960) at 1 hour 39 minutes–1 hour 42 minutes.

References

Akerkar, R., Vega-Gorgojo, G., Løvoll, G., Grumbach, S., Faravelon, A., Finn, R., et al. (2015). *Understanding and mapping big data* (Deliverable D1.1) (pp. 1–101). BYTE

Project, Seventh Framework Programme for ICT. Retrieved from http://byte-project.eu/wp-content/uploads/2016/03/BYTE-D1.1-FINAL-post-Y1-review.compressed-1.pdf.

Aletras, N., Tsarapatsanis, D., Preoţiuc-Pietro, D., & Lampos, V. (2016). Predicting judicial decisions of the European Court of Human Rights: A natural language processing perspective. *PeerJ Computer Science*, *2*, 1–93.

Amoore, L. (2014). Security and the incalculable. *Security Dialogue*, *45*(5), 423–439.

Andrejevic, M. (2013). *Infoglut: How too much information is changing the way we think and know* (1st edn). Abingdon, Oxon ; New York, NY: Routledge.

Angwin, J., Larson, J., Mattu, S., & Kirchner, L. (2016, 23 May). Machine bias: There's software used across the country to predict future criminals. And it's biased against blacks. *ProPublica*. Retrieved from www.propublica.org/article/machine-bias-risk-assessments-in-criminal-sentencing.

Article 29 Working Party. (2014). Statement of the WP29 on the impact of the development of big data on the protection of individuals with regard to the processing of their personal data in the EU, Adopted on 16 September 2014, 14/EN, WP 221.

Ashley, K. D., & Brüninghaus, S. (2006). Computer models for legal prediction. *Jurimetrics*, *46*(3), 309–352.

Ball, K., & Snider, L. (Eds.). (2013). *The surveillance-industrial complex: A political economy of surveillance*. Abingdon, Oxon ; New York, NY: Routledge.

Barry-Jester, A. M., Casselman, B., & Goldstein, D. (2015, 4 August). The new Science of sentencing. *The Marshall Project*. Retrieved from www.themarshallproject.org/2015/08/04/the-new-science-of-sentencing.

Beck, C., & McCue, C. (2009, November). Predictive policing: What can we learn from Wal-Mart and Amazon about fighting crime in a recession? *The Police Chief Magazine*, *76*(11). Retrieved from http://acmcst373ethics.weebly.com/uploads/2/9/6/2/29626713/police-chief-magazine.pdf.

Beck, U. (2002). The terrorist threat world risk society revisited. *Theory, Culture & Society*, *19*(4), 39–55.

Beirne, P. (1987). Adolphe Quetelet and the origins of positivist criminology. *American Journal of Sociology*, *92*(5), 1140–1169.

Berk, R. (2008). Forecasting methods in crime and justice. *Annual Review of Law and Social Science*, *4*(1), 219–238.

Burris, S. (2016, 15 January). "Minority report" is coming true: We now have threat scores to match our credit scores. *Salon*. Retrieved from www.salon.com/2016/01/15/minority_report_is_coming_true_we_now_have_threat_scores_to_match_our_credit_scores_partner/.

Buttarelli, G. (2015). *European data protection supervisor encourages a new debate on big data*. EDPS/2015/11, Brussels, 19 November. Retrieved from https://edps.europa.eu/sites/edp/files/edpsweb_press_releases/edps-2015-11-edps_big_data_en.pdf.

Captain, S. (2015). Hitachi says it can predict crimes before they happen. *Fast Company*. Retrieved from www.fastcompany.com/3051578/elasticity/hitachi-says-it-can-predict-crimes-before-they-happen.

Danziger, S., Levav, J., & Avnaim-Pesso, L. (2011). Extraneous factors in judicial decisions. *Proceedings of the National Academy of Sciences*, *108*(17), 6889–6892.

Davies, W. (2017, 19 January). How statistics lost their power – and why we should fear what comes next. *Guardian*. Retrieved from www.theguardian.com/politics/2017/jan/19/crisis-of-statistics-big-data-democracy.

Deleuze, G. (1992). Postscript on the societies of control. *October*, *59*, 3–7.

Desrosières, A. (2002). *The politics of large numbers: A history of statistical reasoning.* Cambridge, MA: Harvard University Press.

Dewan, S. (2015, 26 June). Judges replacing conjecture with formula for bail. *New York Times*. Retrieved from www.nytimes.com/2015/06/27/us/turning-the-granting-of-bail-into-a-science.html.

European Parliament (2017). European Parliament Resolution of 14 March 2017 on fundamental rights implications of big data: Privacy, data protection, non-discrimination, security and law-enforcement (2016/2225(INI)).

Feeley, M. M., & Simon, J. (1992). The new penology: Notes on the emerging strategy of corrections and its implications. *Criminology*, *30*(4), 449–474.

Foucault, M. (1978). About the concept of the "dangerous individual" in 19th-century legal psychiatry. *International Journal of Law and Psychiatry*, *1*(1), 1–18.

Foucault, M. (1990). *Politics, philosophy, culture: Interviews and other writings, 1977–1984*. (L. Kritzman, Ed.) (Revised edn). New York: Routledge.

Foucault, M. (1995). *Discipline & punish: The birth of the prison*. (A. Sheridan, Trans.). New York: Vintage Books.

Franko Aas, K. (2005). *Sentencing in the age of information: From Faust to Macintosh* (1st edn). London: Routledge-Cavendish.

Garland, D. (1985). The criminal and his science. *The British Journal of Criminology*, *25*(2), 109–137.

Garland, D. (1992). Criminological knowledge and its relation to power: Foucault's genealogy and criminology today. *British Journal of Criminology*, *32*(4), 403–422.

Green, B., Horel, T., & Papachristos, A. V. (2017). Modeling contagion through social networks to explain and predict gunshot violence in Chicago, 2006 to 2014. *JAMA Internal Medicine*. Retrieved from http://jamanetwork.com/journals/jamainternalmedicine/fullarticle/2594804.

Haggerty, K. D., & Ericson, R. V. (2000). The surveillant assemblage. *The British Journal of Sociology*, *51*(4), 605–622.

Harcourt, B. E. (2015). Risk as a proxy for race. *Federal Sentencing Reporter*, *27*(4), 237–243.

Hardoon, D. (2017). *An economy for the 99%: It's time to build a human economy that benefits everyone, not just the privileged few*. Oxfam. Retrieved from http://hdl.handle.net/10546/620170.

Harris, S. (2014, 29 July). The social laboratory. *Foreign Policy*. Retrieved from http://foreignpolicy.com/2014/07/29/the-social-laboratory/.

Harris, S. (2015). *@War: The rise of the military-internet complex* (Reprint edn). Boston: Eamon Dolan/Mariner Books.

Huda, J. (2016, 23 June). Snapping a picture of your hotel room could help stop human trafficking. Retrieved from http://wgntv.com/2016/06/23/snapping-a-picture-of-your-hotel-room-could-help-stop-human-trafficking/.

IBM. (2011, 27 June). IBM Smarter Planet Leadership Series – Memphis Police [CT004]. Retrieved from www.ibm.com/smarterplanet/us/en/leadership/memphispd/.

The International Working Group on Data Protection in Telecommunications. (2014). Working paper on big data and privacy principles under pressure in the age of big data analytics, 55th Meeting, 5–6 May 2014, Skopje.

Jouvenal, J. (2016, 10 January). The new way police are surveilling you: Calculating your threat "score". *Washington Post*. Retrieved from www.washingtonpost.com/local/public-safety/the-new-way-police-are-surveilling-you-calculating-your-threat-score/2016/01/10/e42bccac-8e15-11e5-baf4-bdf37355da0c_story.html.

Kelion, L. (2014, 29 October). London police trial gang violence "predicting" software. *BBC News*. Retrieved from www.bbc.com/news/technology-29824854.

Kleinberg, J., Lakkaraju, H., Leskovec, J., Ludwig, J., & Mullainathan, S. (2017). *Human decisions and machine predictions* (Working Paper No. 23180). National Bureau of Economic Research. Retrieved from www.nber.org/papers/w23180.

Kramer, A. D. I., Guillory, J. E., & Hancock, J. T. (2014). Experimental evidence of massive-scale emotional contagion through social networks. *Proceedings of the National Academy of Sciences, 111*(24), 8788–8790.

Kramer, S. (2017, 23 February). An algorithm is replacing bail hearings in New Jersey. *Motherboard*. Retrieved from https://motherboard.vice.com/en_us/article/an-algorithm-is-replacing-bail-hearings-in-new-jersey.

Kravets, D. (2014, 30 July). Algorithm predicts US Supreme Court decisions 70% of time. *Ars Technica*. Retrieved from https://arstechnica.com/science/2014/07/algorithm-predicts-us-supreme-court-decisions-70-of-time/.

Labi, N. (2012, February). Misfortune teller. *The Atlantic*. Retrieved from www.theatlantic.com/magazine/archive/2012/01/misfortune-teller/308846/.

Lane, J., Stodden, V., Bender, S., & Nissenbaum, H. (Eds.). (2014). *Privacy, big data, and the public good: Frameworks for engagement* (1. publ.). New York, NY: Cambridge University Press.

Leman-Langlois, S. (Ed.). (2008). *Technocrime: Technology, crime and social control* (1st edn). Cullompton: Willan.

Leskovec, J. (2015). *Zakaj se sodniki motijo*. Presented at the Okrogla miza "Pravo v dobi velikega podatkovja: Ali lahko računalnik sodi bolje kot sodnik?", Ljubljana. Retrieved from http://videolectures.net/okroglamizapravo2015_leskovec_sodniki/.

Liessmann, K. P. (2006). *Theorie der Unbildung: Die Irrtümer der Wissensgesellschaft* (17th edn). Vienna: Paul Zsolnay Verlag.

Marr, B. (2016, 20 December). Big data: The 6th "V" everyone should know about. *Forbes*. Retrieved from www.forbes.com/sites/bernardmarr/2016/12/20/big-data-the-6th-v-everyone-should-know-about/.

Martinson, R. (1974). What works? – Questions and answers about prison reform. *The Public Interest, Spring*, 22–54.

Mathiesen, T. (1997). The viewer society Michel Foucault's "Panopticon" revisited. *Theoretical Criminology, 1*(2), 215–234.

Maybin, S. (2016, 17 October). How maths can get you locked up. *BBC News*. Retrieved from www.bbc.com/news/magazine-37658374.

Mayer-Schönberger, V., & Cukier, K. (2013). *Big data: A revolution that will transform how we live, work, and think* (1st edn). Boston: Eamon Dolan/Houghton Mifflin Harcourt.

Mohler, G. O., Short, M. B., Malinowski, S., Johnson, M., Tita, G. E., Bertozzi, A. L., & Brantingham, P. J. (2015). Randomized controlled field trials of predictive policing. *Journal of the American Statistical Association, 110*(512), 1399–1411.

Morozov, E. (2013). *To save everything, click here: Technology, solutionism, and the urge to fix problems that don't exist*. London: Allen Lane.

Office of the Assistant Attorney General. (2014). *The promise and danger of data analytics in sentencing and corrections policy* (No. Washington, DC 20530). Retrieved from www.justice.gov/sites/default/files/criminal/legacy/2014/08/01/2014annual-letter-final-072814.pdf.

O'Neil, C. (2016). *Weapons of math destruction: How big data increases inequality and threatens democracy*. New York: Crown.

Papachristos, A. V. (2007). *Murder by structure: Dominance relations and the social structure of gang homicide in Chicago* (SSRN Scholarly Paper No. ID 855304). Rochester, NY: Social Science Research Network. Retrieved from http://papers.ssrn.com/abstract=855304.

Pasquale, F. (2015). *The black box society: The secret algorithms that control money and information*. Cambridge: Harvard University Press.

Perry, W.L., McInnis, B., Price, C. C., Smith, S., & Hollywood, J. S. (2013). Predictive policing: The role of crime forecasting in law enforcement operations. Santa Monica, CA: RAND Corporation. Retrieved from www.rand.org/pubs/research_reports/RR233.html.

Plesničar M., M., & Klančnik, A. T. (2015). Sodobne rešitve pri pregonu spolne kriminalitete nad otroki. *Pravna Praksa*, *34*(47), 11–14.

Saunders, J., Hunt, P., & Hollywood, J. S. (2016). Predictions put into practice: A quasi-experimental evaluation of Chicago's predictive policing pilot. *Journal of Experimental Criminology*, 1–25.

Sherman, L. W., & Weisburd, D. (1995). General deterrent effects of police patrol in crime "hot spots": A randomized, controlled trial. *Justice Quarterly*, *12*(4), 625–648.

Skeem, J. L., & Lowenkamp, C. T. (2016). *Risk, race, & recidivism: Predictive bias and disparate impact* (SSRN Scholarly Paper No. ID 2687339). Rochester, NY: Social Science Research Network. Retrieved from http://papers.ssrn.com/abstract=2687339.

Smith IV, J. (2015, 9 November). "Minority report" is real – And it's really reporting minorities. *Mic*. Retrieved from https://mic.com/articles/127739/minority-reports-predictive-policing-technology-is-really-reporting-minorities.

Starr, S. B. (2013). *Evidence-based sentencing and the scientific rationalization of discrimination* (SSRN Scholarly Paper No. ID 2318940). Rochester, NY: Social Science Research Network. Retrieved from http://papers.ssrn.com/abstract=2318940.

Turkle, S. (1995). *Life on the screen: Identity in the age of the internet*. New York: Simon & Schuster.

Wacquant, L. (2009). *Prisons of poverty* (Expanded edn). Minneapolis: University of Minnesota Press.

West, Corporation. (2016, 18 February). Empower first responders with enhanced situational awarness. *West Corporation*. Retrieved from www.west.com/safety-services/public-safety/powerdata/beware/.

The White House (2014, May). *Big data: Seizing opportunities, preserving values*. Retrieved from www.whitehouse.gov/sites/default/files/docs/big_data_privacy_report_may_1_2014.pdf.

Wikipedia (2017). *List of countries by number of mobile phones in use*. Retrieved from https://en.wikipedia.org/wiki/List_of_countries_by_number_of_mobile_phones_in_use.

Youyou, W., Kosinski, M., & Stillwell, D. (2015). Computer-based personality judgments are more accurate than those made by humans. *Proceedings of the National Academy of Sciences*, *112*(4), 1036–1040.

8 Subjectivity, algorithms and the courtroom

Mojca M. Plesničar and Katja Šugman Stubbs

Introduction

People make bad decisions all the time: be it the third slice of cake instead of going jogging; hanging out with our high school friend, who always makes us feel bad; spending too little time with our family; spending money we do not have to buy that new bright screen we really do not need, and so on. Bad decisions are, however, not reserved to our private lives; we constantly make bad decisions in our professional lives as well. The field of law, as delicate as it might be, is no exception.

We like to believe that upon entering the courtroom, the judge is magically able to shed her subjectivity and become a neutral, objective observer, who brings nothing of her own to the final decision. We, of course, know that is not true. Schauer (2010: 103) has rather poetically concluded, that it is 'the judge as a human being and not the judge as judge or the judge as lawyer – that has the greatest explanatory power in accounting for judicial behavior'. Accepting this brings us to a conundrum when trying to reconcile the reality of human subjectivity with some of the basic tenets of criminal law, specifically, the requests for neutrality and impartiality of the judges, and the principle of equality, requiring all cases being treated equal regardless of the judge, time, location and any other circumstance. Absolute neutrality or objectivity can never truly be achieved when the one making the decision is a human being, but is there any other option?

There might be. In the modern day era, our judicial courtrooms have evolved in parallel to technological development: stenographers have been replaced by recorders, thick case files by computer documents, court dockets simplified by electronic datasets, and so on, but decision-making has so far remained largely in the hands of the judge. However, the newest developments, stemming from the progress in our abilities to process large amounts of data, have opened the possibilities to a more technology-aided decision-making process, or perhaps even a technology-centred decision-making.

In this chapter, we will attempt to address some of the issues surrounding introducing the use of Big Data into the criminal justice system. Not all areas of courtroom decision-making have so far been tackled, but decisions on bail or sentences are more and more frequently being reviewed through the lens of algorithms with the aim of freeing them from subjective and unreliable human decisions. We will first explore the notion of subjectivity within the criminal decision-making process, arguing that there is certainly a type of harmful subjectivity at play in legal decision-making, but there is also a type of subjectivity, which we may call constructive. Next, we will briefly consider past attempts to limit subjectivity in some areas of judicial decision-making, either through offering additional support to such decision-making or through limiting the judge's discretion. Finally, we will look at what modern algorithms aim to do to address such subjectivity and consider the potential benefits as well as the pitfalls of using Big Data in the courtroom.

Embracing subjectivity

Subjectivity is an inseparable condition of humanity. Being human means being subjective. As was long ago carefully elaborated by the biggest thinkers such as Plato and Kant and consistently confirmed by modern research, we, human beings, do not perceive reality as it is (objectively), but as we interpret it to be. The way we interpret it to be depends on our referential frame: our psychological map of understanding the world. This map is being drawn from the first day we are born (or even from before birth) until the last moments of our lives. It consists of early experience, parental messages, secondary socialisation processes, social, ethical and political values and beliefs, religion etc. All this blends into a unique, unrepeatable referential frame, which contains beliefs about ourselves, others and the world. Humankind is a jigsaw of billions of different referential frames: not even one alike to another. In this respect, we are always each of us subjective.

Nevertheless, our subjectivity does not only make us unique, it also makes us fallible. As illustrated in the introduction, there is no doubt that human decision-making is flawed. Our subjectivity makes different people judge the same issues very differently, and, perhaps more disappointedly, makes us judge similar situations very differently as well. If we are not able to recognise this sad fact learning from our own experience, than this inherent characteristic becomes obvious once faced with the hard-core scientific facts (Ariely, 2008; Kahneman, 2011).

Legal decision-making is no exception to that universal principle (Kapardis, 2009; Klein & Mitchell, 2010). One might expect that the high levels of specialisation and the detailed structuring of the legal proceedings would be conducive towards an environment less prone to error. The decision to study law, the legal education, experience in legal decision-making, the decision to take the bench, experience in judging etc. should all be able to reduce the amount of mistakes in decision-making. However, these are mostly speculations and there is no proof that they actually work in this respect (Schauer, 2010). Most of the research in the field of the psychology of legal decision-making has been done within the American legal system and largely relates to the behaviour of jurors. Only in the late 1970s, research focused on professional judges (Guthrie, Rachlinski, &

Wistrich, 2007; Kahneman, Slovic, & Tversky, 1982; Kapardis, 2009). Lately there has been increased research in the European context as well, led by German researchers (e.g. Mussweiler, Englich, & Strack, 2012). All these studies largely contradict the above assumptions about a diminished fallibility of decision-making in the legal context and confirm that subjectivity is as much a part of legal reality as it is of common life.

Nevertheless, it would be hard to argue the idea that judges should be impartial and objective. Sometimes it is even emphasised that they should be disinterested (Shaman, 1996) in order for legal disputes to be decided free from bias, prejudice or arbitrary personal inclinations. Such demands are very frequent in different legal acts and judicial codes of ethics (e.g. Code of Judicial Conduct, Model Code of Judicial Conduct, Codex of Ethics of Slovenian Judges' Society), but it is also obvious that such a perfect state is impossible to achieve; it is but an ideal. Shaman (1996: 605) puts it nicely when saying 'Pure impartiality is an ideal that can never be completely attained. Judges, after all, are human beings who come to the bench with feelings, knowledge, and beliefs that cannot be magically extirpated.'

However, even if we take such ideals for granted, what do they actually mean? If it is relatively clear what it means that judges have to be impartial, it is much less clear what it means for a judge to be objective? Does it mean being completely detached from life, people and emotions? Does it mean being exclusively rational and capable of excluding emotions? Does it mean being free of values? According to his biographer Supreme Court Justice Oliver Wendell Holmes was close to this 'ideal'.

> Detachment seems the most accurate term to characterize Holmes' stance on the Supreme Court. He was not merely sceptical; his emotions were for the most part not engaged. To put it more precisely, his emotions were stimulated by the professional features of his work but not by its substance.… Acquaintances of Holmes had from his early years noted his apparent indifference to others. His father thought he 'look[ed] at life as at a solemn show where he is only a spectator'.
>
> (White, 1994: 42, quoted from Shaman, 1996: 610)

Should judges therefore be like Mr Spock from *Star Trek*?

There are many different theories on what a judge should be like or what a judicial function encompasses. It is beyond the scope of this chapter to even begin enumerating them. However, one thing seems clear to us: judges have to consider values. As Supreme Justice Cardozo put it: 'The final cause of the law is the welfare of society, and the business of judges is to promote social welfare' (Cardozo, 1921: 66–67, quoted from Shaman, 1996: 617). Judging is about weighing values and considering outcomes of the case regarding the values of society. Only humans hold values. The question therefore is, which are the personal values compatible with the judicial function and able to strengthen it, and which are not. To rephrase the question: is it possible to see some elements of subjectivity as actually helping judges to better exercise their functions? And if this were true, is it not this element of humanity, which, despite being fallible, constitutes the essence of the judicial function? In order to analyse these distinctions therefore one has to distinguish between different aspects of subjectivity. But let us first see what subjectivity means.

The Cambridge online dictionary (2016) defines the word subjective as 'Influenced by or based on personal beliefs or feelings, rather than based on facts'. The Oxford English Dictionary (2016) is more exact in one of its definitions: 'The quality or condition of viewing things chiefly or exclusively through the medium of one's own mind or individuality; the condition of being dominated by or absorbed in one's personal feelings, thoughts, concerns, etc.; individuality, personality.' Solomon (2005) offers one of the more philosophical definitions of subjectivity:

> Subjectivity refers to a person's perspective or opinion, particular feelings, beliefs, and desires. It is often used casually to refer to unsubstantiated personal opinions, in contrast to knowledge and fact-based beliefs. In philosophy, the term is often contrasted with objectivity.

Subjectivity therefore seems something completely contrary to what is required of a judge, since it is the opposite of objectivity and impartiality and even detachment. It is possible to extract from the definitions of subjectivity what seems especially dangerous for the exercise of objectivity: personal beliefs, feelings and desires. True, people have all sorts of beliefs, feelings and desires, some of them completely unreasonable, even insane from the point of view of the social consensus on certain facts. There are, for example, still people who seriously believe that the earth is flat and find ways to produce many 'proofs' for that assertion.[1] If the consensus cannot be reached on such scientifically proven facts, just imagine how much we then differ regarding subtle explanations on human behaviour and reactions.

However, first, most of those beliefs and feelings are never purely individual: each of us grew up in certain social circles and most of our even most personal beliefs are moulded by that experience and share a common root with our reference group (McGarty, Yzebyt, & Spears, 2002). Personal beliefs are therefore never completely personal and unique, even if they are unfounded and unreasonable. Even stereotypes and prejudice, one of the most harmful, yet inevitable creations of the human mind regarding objectivity, are shared by larger groups of people, otherwise they would not be of interest for social psychology but rather for psychopathology. Any given society therefore shares the responsibility for all those beliefs and feelings of its members.

Second, and more importantly, are all those personal beliefs, feelings and desires comprising judicial subjectivity really fatal and harmful for judicial objectivity? Is a judge, for example, who lets her empathy, 'created through the medium of her own mind' for an accused who is a single mother (if we use a stereotypical example) influence her decision while deciding on bail, really acting contrary to her judicial duty? Is a judge who is angry with the defendant a bad judge? The question can also be reversed: are objectivity and detachment (even if achievable) in legal decision-making always and necessarily a good thing? As Yosal Rogat remarked on Supreme Justice Holmes:

> But Holmes more strikingly than any other judge invites a question that is rarely asked: Whether a minimum of involvement is not also required. Holmes was certainly sufficiently detached. Was he, however, sufficiently engaged?
>
> (Rogat, 1964: 243)

The correct answer in our opinion is: it depends. The crucial factor, which makes this subjectivity adequate on the one hand and objectionable on the other hand is the value, which stands behind a certain personal inclination. Maroney taking judicial anger as an example, claims that a good reason for judicial anger is 'one that is accurate, relevant, and reflects good values' (Maroney, 2012: 1250). If this value is consistent with the values promoted by a certain legal system, then it is not only acceptable to be subjective, but also desirable. In order to understand the underlying value behind a personal belief or emotion we need to reconstruct the internal logic of the judge. This might prove to be difficult if not unachievable not only because of lack of time and resources to study such beliefs but also because such internal logic is frequently subconscious and even the judges are not aware of it (Banaji & Greenwald, 2013). Sometimes it is easy to read such underlying thoughts and values from the judgement, sometimes judges even pronounce them, but most commonly, they do not: they just influence their decisions in a certain way.

It is important to understand that subjective emotions, beliefs, thoughts and values are not bad in themselves, but are harmful only if deriving from certain values and beliefs which have no place in a judicial framework, but can be beneficial if they are rooted in others. We therefore have to start distinguishing between different aspects of subjectivity. In our opinion, there are at least two kinds of subjectivities regarding the evaluation of the outcome of judges' decision-making. The first ones are harmful and should never have found their way into legal reasoning, while the second ones enrich legal decisions because they are compatible with basic legal values.

The obvious examples of the first ones, that is the 'harmful subjectivities' are impartialities or subjectivities, which are unacceptable from the point of view of the principle of equality before law and other fundamental legal values such as fairness of procedure etc. Among these we can find bias based in personal interest, personal bias, based on stereotypes (Stangor, 2000) and prejudice (Allport, 1979) or political sympathies or dislikes (Rachlinski, Johnson, Wistrich, & Guthrie, 2009) or likes.

Personal interest, understood as personal financial gain, help to family and friends, and similar, can also lead to objectionable results. A judge personally hurt or involved with either of the parties or an outcome of a case cannot act objectively and impartially and has to be excluded from the case. A terrible, but clear example of this is the 'kids for cash' scandal, in which two Pennsylvanian judges sent juveniles to a detention centre for minor offences in order to receive payment from the private prison facility (Ecenbarger, 2012). It is obvious to any onlooker that such bias is absolutely unacceptable in the judicial, and many other, context.

Similarly there is vast criminological research showing the impact of equally damaging effects of stereotyping and prejudice to impartiality in criminal justice (Gabbidon, Greene, & Young, 2001; Russel, 2013), and it is obvious that acting on such grounds creates the type of subjectivity which is unacceptable from the point of view of equal protection from the law and basic fairness. A Connecticut study of bail-setting, for example, found that judges in New Haven systematically 'overdeter' black male Hispanic defendants from fleeing after release on bail by setting bail at seemingly unjustified high levels for these groups (Ayres & Waldfogel, 1994: 987, 992).

Disappointingly, a great proportion of society may often share such 'subjectivities' in the form of prejudice. If a judge passes a harsher or milder sentence because of the skin colour or sexual appeal of the defendant then such a decision is obviously contrary to the principle of equality and leads to discrimination. The law should be applied to all people alike, regardless of their racial, sexual or other personal differences, and acting on stereotypes and prejudice negates the basic values of law. The same is true of a judge showing favouritism to a political ally or displaying partiality towards a political opponent (Schanzenbach & Tiller, 2008). Such inclinations can be either conscious or subconscious. In the latter case, it is nearly impossible to control and detect them. Rachlinski et al. (2009) found that judges are not immune to such prejudice and bias, and that they significantly influence their decisions.

On the other end of the spectrum there is the 'welcome subjectivity' filled with purely subjective values, feelings or notions, which can actually improve judicial decision-making and make it fairer, more engaged and more human. Can, for example, judicial anger improve judicial decision-making? It can, if it fulfils certain conditions.

A first precondition for all those subjective reactions is, of course, that they are based on an adequate perception of reality and not on some imaginary wrongdoing, distorted perceptions of reality, etc. (Milivojević, 2008: 143–153). Second, the value behind an emotion has to benefit the fairness of the trial. If judicial anger with the defendant's cruel behaviour triggers in a judge a wish for conducting a trial in a more engaged way and a desire to pass a just and even severe sentence, there is nothing wrong with that. Similarly, if judicial anger is triggered by a defence lawyer producing fraudulent evidence, than such judicial anger is more than justified. On the other hand, if judicial anger is triggered by the fact that the defendant reminds the judge of his hated father, or by the fact that the defendant is not of the right race or sex, or if the judge is annoyed with the defendant attorney's aggressive behaviour and wants to punish a defendant more harshly because of that, this is not only unprofessional, but also harmfully subjectively biased. Consequently, if the value behind the emotion is not in accordance with the values promoted by a certain legal system, and if such a subjective reaction remains uncorrected by the judge herself, than the emotion will be inadequate and as a result, the decision based on such an emotional reaction will most likely be flawed as well (Milivojević, 2008: 143–153). If, on the other hand, the emotion triggered is in accordance with the values promoted by a certain legal system, then it can enrich legal decision-making and lead towards a fairer process and outcome.

Third, even an adequate emotion has to be adequately expressed. This means that it has to result in a decision, which is inside of legally defined boundaries, and has to be expressed in a socially acceptable way. Maroney (2012: 1261) referring to Aristotle calls this 'being angry in the right way.' When therefore assessing the adequacy of a certain private emotion (in our case of a judge) according to Milivojević (2008: 128), one has to ask the crucial question of whether an emotion is adequate to the stimuli situation. An adequate emotional reaction is defined as the one which according to its quality (sort of emotion), intensity, length and the manner in which it is expressed corresponds to the objective stimuli situation. If the judge therefore feels strong anger and outrage when confronted with a cruel murder, such a subjective reaction is more than normal if it is adequately expressed and conscious. If, on the other hand, a judge would not feel anything, or would feel these emotions only regarding certain types of defendants or express the emotions by shouting, threatening or passing harsher sentences than legally allowed, than such a subjectivity would not be adequate.

Additionally, in order to understand whether judicial anger is acceptable or not one has to distinguish between a judicial anger which reflects beliefs and values that are worthy of a judge in a democratic society and the ones that are not (Maroney, 2012). When discussing sentencing in Germany, Streng (2007: 154) mentions a similar notion in the context of democratic criminal law when explaining that:

> the judge himself acts as a citizen when determining the punishment, who reflects society's values when assessing the appropriate punishment, whilst keeping within the statutory boundaries. In contrast to a technocratic or an authoritarian criminal law system the judge in our law system relies on values which are coined by his social and professional personality.

However, such distinctions of positive and negative subjectivities have hardly found their room in the legal reality. The 'welcome subjectivity' can easily go unnoticed as, if acting properly, perfectly complements the legal reasoning. Nevertheless, systems never operate in clear-cut hierarchies of values and in making judicial decisions, weighing competing values is a common endeavour. Depending on whether our personal values coincide with those that the judge perceives as leading in a specific case we will appraise her decision as beneficially subjective or as harmfully subjective. In view of such a complex (yet necessary?) assessing of different subjectivities, most systems have opted for (superficially) banning subjectivity altogether.

Avoiding subjectivity

Let's take a closer look at the idea that judges should be prevented from injecting criminal proceedings with their (problematic) subjectivity in the context of sentencing, the criminal phase in which subjectivity has arguably the most room to foster. That has perhaps more to do with the nature of the decision itself than with a conscious systemic determination of the question. The decision on the sentence is charged with meaning that goes far beyond the narrow field of (criminal) law and encompasses philosophical, sociological, penological, criminological and many other considerations. Many contemporary authors agree that the decision on the sentence requires something 'more'. While it is not quite clear what the 'more' is, it certainly is something more than pure rationality (Ashworth, 2010; Henham, 2012; Šelih, 1990).

This mixed nature of sentencing decisions brings about principled as well as practical issues. First, it ties into the long-standing debate on whether the judge's decision is more a product of rational thinking or intuition (Guthrie, Rachlinski, & Wistrich, 2007; Kapardis, 2009; Lovegrove, 2006; Schauer, 2010). On the one hand, according to a formalistic model of legal decision-making based on rationality and analytical thinking, decision-making is a strictly logical operation. It encompasses syllogistic reasoning and does not go beyond that. On the other hand, the realistic model of legal decision-making stresses that every decision, even if within legal boundaries, is a product of an intuitive insight, where a 'legal feeling' is the one guiding the judge. That intuitive decision is only in turn followed by a rationalisation process as a means to explain the decision to the judge herself as well as to a broader audience. The decision is thus 'a kind of intuitive stare, an immediate presence, and even though we used reason to reach it, it is not the same as reason itself' (Furlan, 2002: 163; translation M.M.P.). To an extent this emanates Herbert Simon's idea about intuition being really a sort of recognition (Kahneman, 2011) and 'legal feeling' being intuition rooted in expertise. Some modern scholars have proposed a middle way emphasising the dual nature of the decision on sentencing. According to them, we mostly witness intuitive decision-making, but in some instances intuition can be overridden by rational reflection (Guthrie et al., 2007), thus combining Kahneman's (2011) Systems 1 and 2 of decision-making, the first being fast intuitive thinking and the second slow and deliberate thinking.[2]

Second, the decision on the sentence needs to strike a balance between two (seemingly) competing basic tenets: the principle of equality – requiring all cases to be treated equal, and the concept of individualisation – the tailoring of the sentence to the circumstances of the crime. The latter has largely fallen into disrepute in common law systems at the theoretical level, but still carries a lot of weight in practice. Continental systems typically recognise it at both levels (Frase, 2001; Plesničar, 2013b). Krasnostein and Frieberg (2013) address the same dilemma using the terms 'consistency' and 'individualised justice', which may be more familiar in a common law context. Regardless of the words we use, the predicament is hardly new and has been discussed in different historic settings (e.g. Radzinowicz & Hood [1979] for the common law context, and Bavcon, Vodopivec, & Uderman [1968] for the continental one).

Third, but connected to the previous issue, debates surrounding sentencing often linger on the question of how much discretion should judges be left with in determining the appropriate sentence. Too much discretion allows all sorts of subjectivity to partake in the decision-making process and potentially leads to arbitrariness and disparities. Contrarily, too little discretion leaves no room for subjectivity or the individualisation of the sentence. In practice, different legal systems have opted for different solutions to this conundrum – from broad statutory sentencing ranges to more defined sentencing guidelines and to highly restrictive sentencing tables (see e.g. Ashworth, 2010).

One of the main and cross-jurisdictional tools in searching for an optimal middle ground between too much and too little discretion is the use of case law. Case law has different meanings in common and continental systems as precedents play a much more important role in the former ones. Nevertheless, established case law and its comprehension factor greatly in both types of systems, mainly through reinforcing the principle of equality (Štajnpihler Božič, 2015).

However, searching for relevant case law may sometimes be difficult. First, the criteria used to search for relevant case law are of tremendous importance. The search might only include basic incriminations, but for it to be really useful, it should include subtler notions such as proportionality, aggravating and mitigating circumstances, and other issues. When trying to determine an appropriate sentence in a case of grand larceny, for example, the judge may want to consult previous cases of grand larceny, but not merely in general. Her case involves an aggravating circumstance of a criminal history as well as a mitigating circumstance of the defendant caring for young children (and potentially many others) – she is thus interested in previously decided cases with similar characteristics (Plesničar, 2013a).

Second, the method by which a judge goes about searching relevant case law is vital. In common law systems, where case law is of tremendous importance, unofficial and official year books and reports emerged to help judges search for relevant previous decisions. Even though they served their purpose, such collections are not terribly practical, especially considering the second type of broader searches that we described a judge might need to consider. This may be one of the reasons why technology first entered the decision-making platform in this specific area (Hanson, 2002).

Room for technology?

Different systems improving the relevance and accessibility of case law in the case of sentencing have thus been developed in mainly common law systems (Canada, New Zealand, Australia) in the past decades. In early stages, the development of information technology allowed for simple databases of case law, but soon evolved into more sophisticated and filtered databases with advanced search engines (Hutton, 1995; Miller, 2005; Nissan, 2012; Schild, 2010; Tata, 1998; Tata, Wilson, & Hutton, 1996).

Such databases do not provide numerical solutions, rather they gather data on what sentences were passed in the past in similar cases (Franko, 2005). In this respect, they do not differ tremendously from a non-computerised search for case law where judges seek similar cases in their own judicial experience or that of their colleagues – with the added bonus of such computerised searches being able to prove the judges' recollections empirically. Moreover, we do not consider such tools to be dramatically limiting the judges' subjectivity. In fact, they are able to preserve a significant amount of discretion in sentencing by informing the decision-making process rather than offering a definitive outcome (Hutton, 2013; Krasnostein & Freiberg, 2013).

However, the use of such tools is not without dangers. The discussion about their application ties into the first issues surrounding sentencing that we elaborated above, i.e. whether the judge's decision is a result of rational analysis or intuitive knowledge. The advocates and sympathisers of the latter feared that the heightened use of computerised tools in sentencing might lead to the loss of the 'legal sense' that plays such an important role in sentencing (Lovegrove, 1989; Tata et al., 1996). Moreover, as Hanson (2002) argues, such tools fundamentally change the way we search for and understand the concept of 'law'. Hanson believes that print-based research sources allow a comprehensive outlook at law as a hierarchically organised and structured domain with basic principles and subsequent rules, while computer assisted research blurs such a systemic nature of law and makes it appear no different from other areas of common research of a random and sporadic nature.

Nevertheless, the fears may have been premature. In fact, very few such systems are still functioning, with Canada and Scotland, for example, no longer using them at all (Krasnostein & Freiberg, 2013). It turns out that the distrust towards technology is deeply ingrained in judicial chambers and the technology that judges are reluctant to utilise can hardly survive the test of time (Hutton, 1995; Leith, 1998; Scott, 2013; Tata, 2000).

The second – perhaps more influential – manner in which some legal systems have attempted at taming subjectivity in sentencing is the use of very specific sentencing guidelines and tables as well as mandatory minimum sentences. The literature on sentencing tables is ample and fairly clear – they fared reasonably well in some systems and rather disastrously in others, the US federal sentencing guidelines being the stellar example of the latter (Ashworth, 2009; Mauer, 2001; Reitz, 2005; Stith & Cabranes, 1998; Tonry, 1993, 1997, 2009a, 2016).

Sentencing tables are effective in limiting the judges' discretion (the sentencing ranges are extremely limited and judges not rarely express their discontent about that – see e.g. Auch Schultz, 2012; Bennett, 2012; Kane, 2009; Leipold, 2005; Tucker, 2012; Weinstein, 1993), while the question on their effectiveness in eliminating disparity and contributing towards greater equality or consistency is much more disputed and largely refuted (see e.g. Tonry, 2016).[3]

What is more relevant to our discussion is that sentencing guidelines or even tables are not very technically advanced. However, when applied through modern technology the nature of sentencing tables becomes more apparent: there is a non-official 'sentence calculator' available online (US Federal Sentencing Guidelines calculator, 2015) that is designed after the US Sentencing Commission Guidelines Manual, and its simplicity is rather astounding. One only needs to pick the relevant charge from a list of options and tick the relevant boxes, and the programme itself offers the applicable (narrow) sentencing range. There is little or no room for any kind of 'intuitive insight' (Lovegrove, 1989) or any kind of consideration at all, apart from those relevant to determining which boxes to tick and which not.

As Franko (2005:107) insightfully points out: if judges begin to rely excessively on such tools (and some systems, as the US Federals system did until Booker, even require them to do so!), this may lead to a system in which the fate of the defendant is just a result of a mechanical operation, and the defendant just a 'delinquent without a soul'. This is a result of the 'loss of narrative', which ensues when mechanistic rules are applied, but it can also occur if cases are retrospectively observed using databases and applied as newly created abstract rules to present cases. Moreover, it changes the ways in which we construct the 'essence' of law and interpret what law is and what it is not (Franko, 2004).[4]

The limitations put on judges through sentencing grids (and perhaps even more through mandatory sentences, see e.g. Tonry, 2009b) reflect a deeply ingrained distrust of the professional knowledge commonly attributed to judges. There are numerous factors at play, but the yearning for predictability and greater order as well as perhaps genuine aspiration towards more consistent sentencing decisions seem a very reasonable (partial) explanation (Franko, 2005).

Enter: big data[5]

Notwithstanding the different attempts at limiting harmful subjectivity in sentencing (and criminal justice as a whole), the desired outcomes of neutrality, objectivity and equality have not been reached by far (Provine, 1998; Scott, 2010; Tonry, 2016; Wacquant, 2009). Despite our best efforts, our criminal justice systems are still subject to arbitrariness and disparity and in the face of new technological developments new solutions are being sought in criminal justice as well.

Some important attempts have already been made to improve existing imperfections with aid of different technical and mathematical tools, and learning from huge and complex databases known as Big Data. As Bennett Moses and Chan (2014) nicely explain, Big Data allows us to convert quantitative information about correlations and probabilities into real-world actions through its influence over human decisions.

The use of Big Data in an attempt to guide and improve decisions is one of the most notable trends of the last few years (Ayres, 2008; Mayer-Schönberger & Cukier, 2013). However, it has come to mean much more than the technical details behind its functioning. Boyd and Crawford (2012) point out several components of its contemporary construction, listing technology, analysis as well as mythology. The latter emphasises 'the widespread belief that large data sets offer a higher form of intelligence and knowledge that can generate insights that were previously impossible, with the aura of truth, objectivity, and accuracy' (Boyd & Crawford, 2012: 663). The latter, the 'aura of truth, objectivity, and accuracy', is precisely what criminal justice has desperately been searching for.

The whole point of relying on Big Data is to replace human knowledge, experience and intuition with statistical analysis of enormous amounts of data; therefore to avoid inherent human subjectivity which is viewed as flawed, and to base one's decision on more objective, scientific measures and by this to avoid human imperfection, which can lead to wrong decisions. As Jurgenson (2014) writes: 'The advent of Big Data has resurrected the fantasy of a social physics, promising a new data-driven technique for ratifying social facts with sheer algorithmic processing power.'

This has been attempted in the legal arena as well. There have been many intersections between Big Data and law in the various legal fields seeking to utilise the new technology for different types of decisions and more is yet to come (McGinnis & Pearce, 2014). McGinnis and Pearce (2014) and Katz (2013) give excellent comprehensive accounts into what is happening and list various areas where Big Data has entered the legal landscape. *Lex Machina* for example, is one of the more advanced analytic tools, invented with the purpose of predicting an outcome and the cost of intellectual property litigation. Others have developed models that act as a 'crystal ball' and predict the likelihood of settlement and the expected settlement amount for securities fraud class action lawsuits (McShane, Watson, Baker, & Griffith, 2012). One of the earlier projects proved that algorithms were able to better predict the outcomes and individual votes of the US Supreme Court than renowned professionals (Fowler, Johnson, Ii, Jeon, & Wahlbeck, 2007). If, until recently, it was extremely difficult to imagine any computer analytics to decide complex legal questions such as passing a judgment, counselling or guiding the process of mediation etc. (Dalke, 2013), some of these decisions have been tackled with astonishing success. Just recently, a group of scientists developed a system able to predict the decisions of the European Court of Human Rights with surprising accuracy based just on the textual content from the cases (Aletras, Tsarapatsanis, Preoţiuc-Pietro, & Lampos, 2016).

Criminal courtrooms have so far collided with Big Data at three decision-making levels, all soaked in probabilistic thinking, which humans are so bad at (Kahneman, 2011): (1) decisions on bail (Leskovec, 2015; Milgram, 2013); (2) decisions on sentencing (Hall, Calabro, Sourdin, Stranieri, & Zeleznikow, 2005); and (3) decisions on parole (Harcourt, 2007).

What they all have in common is using a large amount of previously decided cases (the literal Big Data) to build a strong algorithm able to predict the best possible answer to the given question in a specific case (i.e. how likely is the offender to reoffend should she be released on bail/parole or given a community sentence). The methods they use to achieve that are at the core of AI development and are termed 'machine learning' (Kononenko & Kukar, 2007). The outcomes they offer are data-driven probabilities of requested instances, i.e. the likelihood of a defendant skipping bail in bail decisions or the likelihood of reoffending in parole decisions.

The clear answers such algorithms are able to produce are very alluring. They bring promises of a fairer system: informed decisions devoid of bias and any kind of subjectivity. In the best-case scenario, there are many potential benefits of bringing together technological accuracy and human empathy: such decisions could be much more accurate and based on a sound analysis of predictive factors (Donnelly, 2005; Milgram, 2013). Seen as more objective, such algorithms could instil the long-lost trust of the public in the fairness of the criminal justice system (Roberts & Plesničar, 2015). Moreover, they may present an opportunity to purposefully reshape the penal system in order to reflect progressive values and support a more humane outlook.[6]

However, not all is rosy in the land of algorithms. Designing such systems is a tremendously complex undertaking. The premise of machine learning is that the algorithm builds a model for making data-driven predictions not according to a set of pre-imposed rules, but rather through a number of example inputs. This relieves the user from having to set extremely detailed and complex abstract rules, she only has to provide adequate training samples (i.e. previously decided cases). Nonetheless, this is not an easy task in itself. There are important limitations in the number and quality of such training samples. The number needs to be large enough and the variety great enough to build a reliable model that can make adequate generalisations that go beyond the training samples. By increasing the complexity of the problem, the number of different factors (features, attributes) that has to be taken into account increases significantly, which, in turn, requires a huge number of training samples. This means that choosing an appropriate set of training samples as well as choosing the different factors according to which the information is processed, are of tremendous importance (Barocas & Selbst, 2016; Hardt, 2014; Plesničar & Skočaj, 2016).

One of the more prominent examples of that in criminal justice decisions has been the question of race. While it has long been agreed that race should not be a factor in determining probabilistic decisions in criminal justice (see e.g. Harcourt, 2010), the problem has resurfaced raucously in recent debates on the performance of algorithms. ProPublica, the award-winning investigative journalism platform, recently published resonating research into the performance of COMPAS, an algorithm widely used in all three types of decisions described earlier across the US. The authors accused COMPAS of being racially biased and labelling black defendants as likely to reoffend within the next two years at almost twice the rate of white defendants. Conversely, white defendants were labelled as low risk more often than black defendants were (Angwin, Larson, Mattu, & Kirchner 2016). Their findings have been thoroughly scrutinised by different professionals (ironically, much more than a typical algorithm being implemented in criminal justice!) and refuted by COMPAS's author Northpointe (Dietrerich, Mendoza, & Brennan, 2016). The issue relevantly ties into the ongoing debate about whether modern risk predictors such as criminal history act as surrogates for racial predictors or not (Hannah-Moffat, 2013; Harcourt, 2010; Skeem & Lowenkamp, 2016).

Nevertheless, even if we accept that race or other problematic demographic determinants are not consciously built-in in the algorithms by programmers, the problem does not disappear. The model of decision-making built through machine learning is not supposed to apply abstract rules on concrete cases. What it does is offer decisions similar to those that occurred in the past (of which it learnt in the training samples). In other words, 'an algorithm is only as good as the data it works with' (Barocas & Selbst, 2016: 671). Earlier we argued that judges instil personal subjectivity into their decisions, subjectivities that often stem from socially sanctioned prejudice and bias. It only follows that algorithms built on such cases and learning from such cases will replicate the same subjectivities, the very ones we were trying to avoid by using them in the first place (Hardt, 2014).

It is conceivable to imagine a perfect training sample made of perfect decisions by excellent judges – but this is not the case in reality: if we are convinced judges cannot be perfect – their decisions cannot be perfect either. However, even if we were able to produce such a perfect set of training samples, it might not be enough. The law is not static; rather it is a very dynamic, ever-evolving notion. Algorithmic models should therefore be able to encompass such changes and be able to adapt to new evidence or new paradigms and do it rather quickly, and there are some concepts in machine learning that give great promises to achieve that in the future (Plesničar & Skočaj, 2016).[7] Yet as of now, the systems being used in actual criminal justice cases are not capable of such things.

This brings us to the question of validity and verification. EPIC (a public interest research centre) gives a good visual imprint into what the state of validation is in the US, where such tools are most commonly used (EPIC, 2016). Out of the 50 states using different types of predictive algorithms, only ten have bothered to conduct validity studies. This does not mean that the models have not been tested yet in any way; on the contrary, companies marketing such models offer assurances about their robustness, only typically choose to not make them thoroughly transparent (Diakopoulos, 2014). Such was, for instance, the case with COMPAS, where Northpointe refused to provide ProPublica with the inner workings of the algorithm despite having published a paper on its effectiveness. In order to check it for bias (which Northpointe had not done before), ProPublica turned to the sheriff of Broward County, Florida and received the required data through a public records request (Gong, 2016). Even if tested by their authors, the robustness and performance they are looking for in the algorithm may not be the one conducive to justice. Moreover, transparency may sometimes actually be hard to achieve, as the operational details of machine learning are not always clear and easy to understand (Tufekci, 2016).

Last, we bring in a question previously analysed with regard to information databases in sentencing. The dehumanising effect such instruments were feared to cause was largely averted by a simple reality: judges did not use them enough for them to 'kick in'. How likely is that to happen with Big Data in criminal justice? Our guess is: not very (cf. Barton, 2013 for a different outlook). On the one hand, there is some lingering aversion towards new types of technology being used in such manner among the legal professionals. Combined with the belief of the judiciary that they are solely responsible for e.g. passing a just sentence, those are factors that supposedly led to the demise of information databases in some jurisdictions. On the other hand, however, the circumstances have changed. Modern day judges have largely been allowed ample time to adjust to new sorts of technologies to a much bigger extent than their counterparts in the 1990s. Furthermore, there is no longer a general distrust towards technology and much less distrust towards Big Data (Katz, 2013; McGinnis & Pearce, 2014). In fact, Big Data is often hyped to be the future of decision-making in very different fields, including criminal justice and is very persuasive in its discourse about objectivity, robustness and reliability (Boyd & Crawford, 2012; Mayer-Schönberger & Cukier, 2013).

So far, the use of Big Data in criminal justice has not been mandatory, at least not in the sense of the decision-maker being bound to follow it. The decisions presented to judges are always probabilistically put (something we may have bigger problems with than we acknowledge – Donnelly, 2005), and the judges are free to deviate from them. However, the psychology behind such deviations may make them less likely: a judge deciding against the prediction of an algorithm is seemingly taking a greater risk and greater responsibility, even though her decision would be the same as without the algorithmic analysis. It does not seem too far-fetched to imagine judges being reluctant to take such additional heat, especially in systems where their mandates are not permanent but subject to popular vote.

Conclusion

In the course of this chapter, we have explored two main issues. On the one hand, the notion of subjectivity, where we have tried to expose its dark side as well as its bright side. In this vein, we have argued that subjectivity in line with the values of the legal system and within its limits is something desirable and valuable.

On the other hand, we have spent some time thinking about different ways in which different systems have attempted to banish harmful subjectivity from legal decision-making, largely as a reflection of a deeply entrenched distrust of professional knowledge. In earlier attempts, especially with sentencing via sentencing tables, virtually all subjectivity has been removed, in turn causing important and grave problems.

Newer approaches, those using Big Data seem more promising; in spite of the many limitations that we have listed. Quite contrary, if we consider all those implications and are able to combine the benefits of Big Data with the benefits of subjectivity, we could perhaps achieve a much fairer system of criminal justice.

However, we do not believe that can be achieved instantly. On the contrary, this should be a deliberately slow process with ample time for analysing, rethinking and contemplating its many implications.

The first, crucial, instance in which Big Data can be invaluable, is simply putting up a mirror to our existing decision-making. Since emulating existing practice is the way algorithms perform, reflecting in the images they present can undress underlying problems – which we know we have many. If algorithms show bias, it is very likely they are imitating existing bias. Building algorithms in ways that would allow us to understand how decisions were reached, instead of leaving them in black boxes, would allow us to recognise faults in our human decision-making more clearly and hopefully amend them.

Second, we do not believe such systems should be implemented right away. Acknowledging imperfections in human decision-making makes it clear that decision-making modelled upon it will not be infallible. Leaving room and time to develop while testing and validating it may seem too prudent and incompatible with modern penchants for efficiency, but to us it seems valuable. This would also allow for further developments in machine learning that could tremendously improve the performances of algorithms, as well as for developments in designing and factoring the appropriate features to be included in such decisions.

Third, Big Data may be able to avoid some of the pitfalls into which sentencing tables have fallen. Subjectivity here is not necessarily omitted, quite the opposite, it is emulated. This means that, yes, harmful subjectivity can be copied, but it also means that welcome subjectivity can be emulated as well. Moreover, sentencing tables needed to diminish the number of relevant factors as much as possible to be operational; algorithms need to do no such thing. They can encompass a much much greater number of circumstances and factors, and solely by doing that allow for much more nuanced decisions than the two-dimensional sentencing grids were ever able to.

Last, however, this does not mean that we believe Big Data could stand on its own. The only reasonable and acceptable way of entangling it into criminal justice is to do so with full awareness of its need to be interpreted. Judges should be able to understand and appreciate the aid of Big Data, but reminded to never mistake probabilistic answers for truths. Removing the human from criminal justice is thoroughly inconceivable, more so in a time when the field desperately needs more empathy and more of the 'human' component.

Notes

1 See http://blackbag.gawker.com/the-earth-is-flat-explained-1755002534 (17 July 2016).

2 Kahneman's System 1 encompasses quick, automatic and emotional responses that are subconscious, while System 2 covers decision-making that is slow, deliberate, and requires effort and attention. Both systems are innate in human decision-making (Kahneman, 2011).

3 Stith and Cabranes (1998) have fittingly pointed out the similarity between the US federal sentencing guidelines and a system of sentencing designed in the early twentieth century by Enrico Ferri (one of Cesare Lombroso's more prominent pupils). Ferri's system contained a precise compilation of sentencing determinants and instructions on penal arithmetic, and was deemed by his contemporaries to be overly mechanistic and complicated, and was as such never implemented, unlike its modern day counterpart. Ferri's system as well as modern day sentencing tables are understood to be unduly simplistic and utterly missing of the element of 'more' with which we begun the conversation on sentencing.

4 This is not an occurrence unique to the field of law. For an example of similar changes in Dutch social work see e.g. Keymolen & Broeders, 2013.
5 We thank Danijel Skočaj for the numerous discussions on the issue and for the willingness and patience to explain the concept and mechanics of machine learning to social scientists.
6 If, for example, the current error rate (the proportion of defendants that do not comply with bail requirements) of judges, when deciding on bail, is 17 per cent, the algorithm could be set to either reduce that error rate to a minimum (Leskovec, 2015), or to retain the 17 per cent error rate, that seems socially acceptable, while letting a much greater proportion of defendants out on bail. See also Hannah-Moffat, 2013.
7 New modes of machine learning (e.g. one- and zero-shot learning, knowledge-transfer etc.) that are being developed could greatly improve the adaptability and openness of algorithmic systems to fast and direct human intervention (Plesničar & Skočaj, 2016).

Bibliography

Aletras, N., Tsarapatsanis, D., Preoţiuc-Pietro, D., & Lampos, V. (2016). Predicting judicial decisions of the European Court of Human Rights: A natural language processing perspective. *PeerJ Computer Science, 2*, e93. Retrieved from https://doi.org/10.7717/peerj-cs.93.

Allport, G. (1979). *The nature of prejudice* (3rd edn). Cambridge, MA: Perseus Books.

Angwin, J., Larson, J., Mattu, S., & Kirchner, L. (2016). Machine bias: There's software used across the country to predict future criminals. And it's biased against blacks. *ProPublica*. Retrieved from www.propublica.org/article/machine-bias-risk-assessments-in-criminal-sentencing

Ariely, D. (2008). *Predictibly irrational: The hidden forces that shape our decisions.* New York: HarperCollins.

Ashworth, A. (2009). Techniques for reducing sentence disparity. In A. von Hirsch, A. Ashworth, & J. V. Roberts (Eds.), *Principled sentencing: Readings on theory and policy* (3rd edn, pp. 241–257). Oxford; Portland: Hart.

Ashworth, A. (2010). *Sentencing and criminal justice* (5th edn). Cambridge: Cambridge University Press.

Auch Schultz, T. (2012). Judge says some child porn sentencing guidelines are too harsh. *Post-Tribune (Chicago-Sun Times)*. Retrieved from http://posttrib.suntimes.com/news/lake/13720352-418/judge-says-some-child-porn-sentencing-guidelines-are-too-harsh.html.

Ayres, I. (2008). *Super crunchers: Why thinking-by-numbers is the new way to be Smart.* New York: Bantam Books.

Ayres, I., & Waldfogel, J. (1994). A market test for race discrimination in bail setting. *Stanford Law Review, 46*, 987–1047.

Banaji, M. R., & Greenwald, A. G. (2013). Blindspot: Hidden biases of good people. New York: Delacorte Press.

Barocas, S., & Selbst, A. D. (2016). Big data's disparate impact. *California Law Review, 104*, 671–732.

Barton, B. H. (2013). The lawyer's monopoly – what goes and what stays. *Fordham Law Review, 82*, 3067.

Bavcon, L., Vodopivec, K., & Uderman, B. (1968). *Individualizacija kazni v praksi naših sodišč*. Ljubljana: Inštitut za kriminologijo pri Pravni fakulteti v Ljubljani.

Bennett, M. W. (2012). How mandatory minimums forced me to send more than 1,000 nonviolent drug offenders to federal prison. *The Nation*. Retrieved from www.thenation.com/article/170815/how-mandatory-minimums-forced-me-send-more-1000-nonviolent-drug-offenders-federal-pri.

Bennett Moses, L., & Chan, J. (2014). Using big data for legal and law enforcement decisions: Testing the new tools. *University of New South Wales Law Journal*, *37*(2), 643–678.

Boyd, D., & Crawford, K. (2012). Critical questions for big data. *Information, Communication & Society*, *15*(5), 662–679. Retrieved from https://doi.org/10.1080/1369118X.2012.678878.

Boyle, J. (1991). Is subjectivity possible? The post-modern subject in legal theory. *University of Colorado Law Review*, 62, 489–524.

Cambridge Dictionary (2016), available at: http://dictionary.cambridge.org/dictionary/english/subjective?q=subjectivity, accessed 17 July 2016.

Cardozo, B. M. (1921). *The nature of the judicial process*. New Haven: Yale University Press.

Dalke, D. L. (2013). Can computers replace lawyers, mediators and judges? *Advocate Vancouver*, 71, 703–710.

Diakopoulos, N. (2014). *Algorithmic accountability reporting: On the investigation of black boxes*. Columbia Journalism School, Tow Centre. Retrieved from http://tow-center.org/wp-content/uploads/2014/02/78524_Tow-Center-Report-WEB-1.pdf.

Dietrerich, W., Mendoza, C., & Brennan, T. (2016). *COMPAS risk scales: Demonstrating accuracy equity and predictive parity – performance of the COMPAS risk scales in Broadward County*. Traverse City, Michigan: Northpointe. Retrieved from www.documentcloud.org/documents/2998391-ProPublica-Commentary-Final-070616.html.

Donnelly, P. (2005). *How juries are fooled by statistics*. Oxford. Retrieved from www.ted.com/talks/peter_donnelly_shows_how_stats_fool_juries.

Ecenbarger, W. (2012). Kids for cash: Two judges, thousands of children, and a $2.6 million kickback scheme. New York: The New Press.

EPIC (2016) available at http://epic.org/algorithmic-transparency/crim-justice/ for a comprehensive grid of instruments used by various states and their testing (accessed 20 November 2016).

Fairfield, J., & Shtein, H. (2014). Big data, big problems: Emerging issues in the ethics of data science and journalism. *Journal of Mass Media Ethics*, *29*(1), 38–51. Retrieved from https://doi.org/10.1080/08900523.2014.863126.

Fowler, J. H., Johnson, T. R., Ii, S. F, J., Jeon, S., & Wahlbeck, P. J. (2007). Network analysis and the law: Measuring the legal importance of supreme court precedents. *Political Analysis*, *15*(3), 324–346.

Franko, K. (2004). From narrative to database Technological change and penal culture. *Punishment & Society*, *6*(4), 379–393. Retrieved from https://doi.org/10.1177/1462474504046119.

Franko, K. (2005). *Sentencing in the age of information: From Faust to Macintosh*. London: Routledge.

Frase, R. S. (2001). Comparative perspectives on sentencing policy and research. In M. H. Tonry, & R. S. Frase (Eds.), *Sentencing and sanctions in western countries* (pp. 259–292). Oxford: Oxford University Press.

Friedman, B., & Nissenbaum, H. (1997). Bias in Computer Systems. In B. Friedman (Ed.), *Human values and the design of computer technology* (pp. 21–40). Cambridge: Cambridge University Press.

Furlan, B. (2002). *Problem realnosti prava*. (M. Pavčnik, Ed.). Ljubljana: Cankarjeva založba.

Gabbidon, S. L., Greene, H. T., & Young, V. D. (Eds.). (2001). *African American classics in criminology and criminal justice*. Thousand Oaks: Sage.

Gong, A. (2016). *Ethics for powerful algorithms (1 of 4) – Abe Gong*. Retrieved 12 November 2016, from https://medium.com/@AbeGong/ethics-for-powerful-algorithms-1-of-3-a060054efd84#.t5ado8k68.

Guthrie, C., Rachlinski, J., & Wistrich, A. (2007). Blinking on the bench: How judges decide cases. *Cornell Law Review*, *93*(1), 1–44.

Hall, M. J. J., Calabro, D., Sourdin, T., Stranieri, A., & Zeleznikow, J. (2005). Supporting discretionary decision-making with information technology: A case study in the criminal sentencing jurisdiction. *University of Ottawa Law & Technology Journal*, *2*, 1–36.

Hannah-Moffat, K. (2013). Actuarial sentencing: An 'unsettled' proposition. *Justice Quarterly*, *30*(2), 270–296. Retrieved from https://doi.org/10.1080/07418825.2012.682603.

Hanson, F. A. (2002). From key numbers to keywords: How automation has transformed the law. *Law Library Journal*, *94*(4), 563–600.

Harcourt, B. E. (2010). Risk as a proxy for race. *Criminology & Public Policy*. Retrieved from https://papers.ssrn.com/abstract=1677654.

Hardt, M. (2014). How big data is unfair. Retrieved 11 November 2016, from https://medium.com/@mrtz/how-big-data-is-unfair-9aa544d739de#.nx6ofxaji.

Henham, R. J. (2012). *Sentencing and the legitimacy of trial justice*. Abingdon, New York: Routledge.

Hutton, N. (1995). Sentencing, rationality, and computer technology. *Journal of Law and Society*, *22*(4), 549–570. Retrieved from https://doi.org/10.2307/1410614.

Hutton, N. (2013). From intuition to database: Translating justice. *Theoretical Criminology*, *17*(1), 109–128. Retrieved from https://doi.org/10.1177/1362480612465767.

Jurgenson, N. (2014). View From nowhere. Retrieved 19 September 2016, from http://thenewinquiry.com/essays/view-from-nowhere/.

Kahneman, D. (2011). *Thinking, fast and slow*. New York: Farrar, Straus and Giroux.

Kahneman, D., Slovic, P., & Tversky, A. (1982). *Judgment under uncertainty: Heuristics and biases*. Cambridge: Cambridge University Press.

Kane, J. L. *United States* v. *John Brownfield*, Jr., Memorandum opinion and order on sentencing, No. Criminal Case No. 08-cr-00452-JLK (US District Court of Colorado December 2009). Retrieved from www.scribd.com/doc/92111409/Judge-Kane-Sentencing-Memo.

Kapardis, A. (2009). *Psychology and law: A critical introduction*. Cambridge: Cambridge University Press.

Katz, D. M. (2013). Quantitative legal prediction or how I learned to stop worrying and start preparing for the data-driven future of the legal services industry. *Emory Law Journal*, *62*. Retrieved from https://works.bepress.com/daniel_m_katz/14/.

Keymolen, E., & Broeders, D. (2013). Innocence lost: Care and control in Dutch digital youth care. *British Journal of Social Work*, *43*(1), 41–63. Retrieved from https://doi.org/10.1093/bjsw/bcr169.

Klein, D. E., & Mitchell, G. (2010). *The psychology of judicial decision making*. Oxford: Oxford University Press.

Kononenko, I., & Kukar, M. (2007). *Machine learning and data mining*. Chichester, UK: Woodhead Publishing.

Kranzberg, M. (1986). Technology and history: 'Kranzberg's Laws'. *Technology and Culture*, *27*(3), 544–560.

Krasnostein, S., & Freiberg, A. (2013). Pursuing consistency in an individualistic sentencing framework: If you know where you're going, how do you know when you've got there? *Law and Contemporary Problems*, *76*(1), 265–288.

Leipold, A. D. (2005). Why are federal judges so acquittal prone? *Washington University Law Quarterly, 83*(1), 151–227.

Leith, P. (1998). The judge and the computer: How best decision support? *Artificial Intelligence and Law, 6*(2), 289–309. Retrieved from https://doi.org/10.1007/978-94-015-9010-5_6.

Leskovec, J. (2015). *Zakaj se sodniki motijo*. Ljubljana. Retrieved from http://videolectures.net/okroglamizapravo2015_leskovec_sodniki/.

Lovegrove, A. (1989). *Judicial decision making, sentencing policy and numerical guidance*. New York: Springer.

Lovegrove, A. (2006). *The framework of judicial sentencing: A study in legal decision making*. Cambridge: Cambridge University Press.

Maroney, T. A. (2012). Angry judges. *Vanderbilt Law Review, 65*, 1207–1286.

Mauer, M. (2001). The causes and consequences of prison growth in the United States. *Punishment & Society, 3*(1), 9–20. Retrieved from https://doi.org/10.1177/14624740122228212.

Mayer-Schönberger, V., & Cukier, K. (2013). *Big data: A revolution that will transform how we live, work, and think*. Boston; New York: Houghton Mifflin Harcourt.

McGarty, C., Yzerbyt, V. Y., & Spears, R. (2002). Social, cultural and cognitive factors in stereotype formation. In C. McGarty, V. Y. Yzerbyt, & R. Spears, (Eds.), *Stereotypes as explanations* (pp. 1–15). Cambridge: Cambridge University Press.

McGinnis, J. O., & Pearce, R. G. (2014). The great disruption: How machine intelligence will transform the role of lawyers in the delivery of legal services. *Fordham Law Review, 82*, 3041–3066.

McShane, B., Watson, O., Baker, T., & Griffith, S. (2012). Predicting securities fraud settlements and amounts: A hierarchical bayesian model of federal securities class action lawsuits. *Faculty Scholarship*. Retrieved from http://scholarship.law.upenn.edu/faculty_scholarship/409.

Milgram, A. (2013). *Why smart statistics are the key to fighting crime*. San Francisco. Retrieved from www.ted.com/talks/anne_milgram_why_smart_statistics_are_the_key_to_fighting_crime.

Milivojević, Z. (2008). *Emocije: razumevanje čustev v psihoterapiji*. Novi Sad: Psihopolis institut.

Miller, M. L. (2005). A map of sentencing and a compass for judges: Sentencing information systems, transparency, and the next generation of reform. *Columbia Law Review, 105*(4), 1351–1395.

Mussweiler, T., Englich, B., & Strack, F. (2012). Anchoring effect. In R. F. Pohl (Ed.), *Cognitive illusions: A handbook on fallacies and biases in thinking, judgement and memory* (pp. 183–200). London: Psychology Press.

Nissan, E. (2012). *Computer applications for handling legal evidence, police investigation and case argumentation*. New York: Springer.

O'Neil, C. (2016). *Weapons of math destruction: How big data increases inequality and threatens democracy*. New York: Crown.

Oxford Dictionaries (2016), available at: www.oed.com/view/Entry/192707?redirectedFrom=subjectivity#eid, accessed 17 July 2016.

Plesničar, M. M. (2013a). Odločanje o sankcijah v kontinentalnih pravnih ureditvah. In M. Ambrož, K. Filipčič, & A. Završnik (Eds.), *Zbornik za Alenko Šelih: kazensko pravo, kriminologija, človekove pravice* (pp. 326–330). Ljubljana: Inštitut za kriminologijo pri Pravni fakulteti v Ljubljani, Pravna fakulteta, Slovenska akademija znanosti in umetnosti.

Plesničar, M. M. (2013b). The individualization of punishment: Sentencing in Slovenia. *European Journal of Criminology*, *10*(4), 462–478.

Plesničar, M. M., & Skočaj, D. (2016). The predicament of decision-making in machines and humans. In *Machine ethics and machine law, e-proceedings*. Krakow: Jagiellonian University Krakow. Retrieved from http://machinelaw.philosophyinscience.com/wp-content/uploads/2016/06/PROCEEDINGS-ver1-1.pdf.

Posner, E. (2008). Does political bias in the judiciary matter? Implications of judicial bias studies for legal and constitutional reform. *The University of Chicago Law Review*, 75, 853–883.

Provine, D. M. (1998). Too many black men: The sentencing judge's dilemma. *Law & Social Inquiry*, *23*(4), 823–856.

Rachlinski, J. J., Johnson, S. L., Wistrich, A. J., & Guthriet, C. (2009). Does unconscious racial bias affect trial judges? *Notre Dame Law Review*, *84*(3), 1195–1246.

Radzinowicz, L., & Hood, R. (1979). Judicial discretion and sentencing standards: Victorian attempts to solve a perennial problem. *University of Pennsylvania Law Review*, *127*, 1288–1349.

Reitz, K. R. (2005). The new sentencing conundrum: Policy and constitutional law at cross-purposes. *Columbia Law Review*, *105*(4), 1082–1123.

Roberts, J. V., & Plesničar, M. M. (2015). Sentencing, legitimacy, and public opinion. In G. Meško, & J. Tankebe (Eds.), *Trust and legitimacy in criminal justice* (pp. 33–51). Springer International Publishing. Retrieved from http://link.springer.com/chapter/10.1007/978-3-319-09813-5_2.

Rogat Y. (1964). The judge as spectator. *University of Chicago Law Review*, *31*(2), 213–256.

Russel B. (Ed.). (2013). *Perception of female offenders: How stereotypes and social norms affect criminal justice responses*. New York: Springer.

Schanzenbach, M. M., & Tiller, E. H. (2008). Reviewing the sentencing guidelines: Judicial politics, empirical evidence, and reform. *University of Chicago Law Review*, *75*(2), 715–760.

Schauer, F. (2010). Is there a psychology of judging? In D. E. Klein, & G. Mitchell (Eds.), *The psychology of judicial decision making* (pp. 103–120). Oxford: Oxford University Press.

Schild, U. J. (2010). Criminal sentencing and intelligent decision support. In G. Sartor, & L. K. Branting (Eds.), *Judicial applications of artificial intelligence* (pp. 47–98). Dordrecht: Kluwer Academic Publishers.

Scott, R. W. (2010). Inter-judge sentencing disparity after Booker: A first look. *Stanford Law Review*, *63*(1), 1–55.

Scott, R. W. (2013). The skeptic's guide to information sharing at sentencing. *Utah Law Review*. Retrieved from https://ssrn.com/abstract=2160484.

Shaman, J. M. (1996). The impartial judge: Detachment of passion? *DePaul Law Review*, *45*(3), 605–632.

Šelih, A. (1990). *Sodna odmera kazni*. Ljubljana: Inštitut za kriminologijo pri Pravni fakulteti v Ljubljani.

Skeem, J. L., & Lowenkamp, C. T. (2016). Risk, race, and recidivism: Predictive bias and disparate impact. *Criminology*. Retrieved from https://doi.org/10.1111/1745-9125.12123.

Solomon, R. C. (2005). *Subjectivity* (T. Honderich, Ed.). Oxford Companion to Philosophy. Oxford: Oxford University Press.

Štajnpihler Božič, T. (2015). The role of case law in judicial decision-making : A sociological perspective. *Sociologija*, *57*(4), 593–619

Stangor, C. (Ed.). (2000). *Stereotypes and prejudice: Essential readings*. Philadelphia: Taylor & Francis.

Stith, K., & Cabranes, J. A. (1998). *Fear of judging: Sentencing guidelines in the federal courts*. Chicago: University of Chicago Press.

Streng, F. (2007). Sentencing in Germany: Basic questions and new developments. *German Law Journal*, *8*(2), 153–171.

Tata, C. (1998). The application of judicial intelligence and 'rules' to systems supporting discretionary judicial decision-making. *Artificial Intelligence and Law*, *6*(2–4), 203–230. Retrieved from https://doi.org/10.1023/A:1008274209036.

Tata, C. (2000). Resolute ambivalence: Why judiciaries do not institutionalize their decision support systems. *International Review of Law, Computers & Technology*, *14*(3), 297–316. Retrieved from https://doi.org/10.1080/713673373.

Tata, C., Wilson, J., & Hutton, N. (1996). Representations of knowledge and discretionary decision-making by decision-support systems: The case of judicial sentencing. *Journal of Information, Law and Technology*, *1*(2). Retrieved from www2.warwick.ac.uk/fac/soc/law/elj/jilt/1996_2/tata.

Tonry, M. H. (1993). The failure of the US Sentencing Commission's guidelines. *Crime & Delinquency*, *39*(2), 131–149. Retrieved from https://doi.org/10.1177/0011128793039002001.

Tonry, M. H. (1997). *Sentencing matters*. Oxford; New York: Oxford University Press.

Tonry, M. H. (2009a). Explanations of American punishment policies: A national history. *Punishment Society*, *11*(3), 377–394. Retrieved from https://doi.org/10.1177/1462474509334609.

Tonry, M. H. (2009b). The mostly unintended effects of mandatory penalties: Two centuries of consistent findings. *Crime and Justice*, *38*(1), 65–114.

Tonry, M. H. (2016). *Sentencing fragments: Penal reform in America, 1975–2025*. Oxford: Oxford University Press.

Tucker, J. H. (2012). How fair are federal sentencing guidelines? One skeptical district judge weighs in. *River Front Times*. Retrieved from http://blogs.riverfronttimes.com/dailyrft/2012/05/judge_kane_john_brownfield_federal_sentencing_guidelines.php.

Tufekci, Z. (2016). *Machine intelligence makes human morals more important*. Banff, Canada. Retrieved from www.ted.com/talks/zeynep_tufekci_machine_intelligence_makes_human_morals_more_important.

US Federal Sentencing Guidelines calculator. (2015). Available at: http://sentencing.us, accessed 25 November 2016.

Wacquant, L. (2009). *Prisons of poverty*. Minneapolis: University of Minnesota Press.

Weinstein, J. B. (1993). Memorandum. *Federal Sentencing Reporter*, *5*(5), 298–298. Retrieved from https://doi.org/10.2307/20639593.

White, G. E. (1994). Intervention and detachment: Essays in legal history and jurisprudence. Oxford: Oxford University Press.

Part V

Big data automation limitations

9 Judicial oversight of the (mass) collection and processing of personal data

Primož Gorkič

Introducing the problem

What makes taking a DNA sample worthy of judicial oversight? It may bring but a light touch on the inside of a person's cheek. Or, it may be even less invasive, for example, by taking a hair sample from a hair brush or analysing a used cup of coffee left behind in your favourite coffee shop. Perhaps it is easier to understand why only a judge is competent to authorise the interception of your telephone calls or e-mails; a sense of being violated when your most intimate conversations are shared with some government official is easier to imagine. In some instances, however, this sense of being violated never comes to play. Perhaps at this very moment, a police officer is analysing a list of Internet protocol (IP) addresses you visited last month, noting your frequent visits to a website belonging to a law office, to a local Alcoholics Anonymous or to a political party. Perhaps during a search for a partial match in a DNA database your name just popped out, a police officer realising a familial connection you could never even imagine.

On a larger scale, the extent of covert surveillance and bulk accumulation of communications (traffic) data is matched only by the amount of facilities built to store and process such acquired data. In order to at least attempt to comprehend the impact of such practices it may be useful to attempt pairing each telephone number you called last month with a name of a person or an organisation. Now, imagine this being done right now by a government official.

The problem at hand is simple: should legislation provide for a form of judicial oversight of such government activities? If the answer in case of interception of communication is a resounding yes, should the same apply to, for example, DNA sampling, testing and processing, or to acquiring, accessing and processing data about communication traffic?

In attempting an answer, this chapter proceeds in the following steps. First, we will examine how different jurisdictions approach such a question. The approaches are sufficiently varied and allow a further exploration into how various courts or legislations substantiate different normative solutions. This will be the second step: why do some jurisdictions require or provide for judicial oversight and some do not? Finally, we need to explore the different concepts that have guided courts in their determination of safeguards in cases of privacy interference. These may, in turn, offer an answer, why should we even bother with judicial oversight: is there any added value in submitting such government practices to judicial oversight? It seems that different approaches may very well rest on a distinctly different understanding of what protecting privacy actually entails.

Is mass data processing worthy of judicial supervision?

Even a brief, cursory look at selected jurisdictions shows that the scope of judicial supervision varies not only across jurisdictions but also with regard to the type of personal data in question.

Communication traffic data

As the pervasive surveillance practices of US government agencies (among others) have become widely known, so have the jurisdictional differences in approaching the protection of so-called communication traffic (meta) data. Within the US jurisdiction, the acquisition of communication traffic data traditionally falls under the so-called third party doctrine (*Smith* v. *Maryland*, 442 US 735 [1979]); see, for example, Kerr, 2009, p. 177 *et seq.*). The importance of the doctrine in justifying the constitutionality of government practices is illustrated by arguments presented in *ACLU* v. *Clapper*, 785 F.3d 787 (2d Cir. 2015). The defendant – the government – essentially argued that individuals are not entitled to the Fourth Amendment protections, after they have voluntarily submitted the relevant data to a third party, e.g. their service provider.

The application of the third party doctrine to communication traffic data is not new, its implications in times of staggering technological advancements, however, are. At this point, it is quite irrelevant, whether the third party doctrine essentially amounts to a waiver of Fourth Amendment Protections (see, e.g. Fabbrini, 2015, p. 91), or whether it results in no protection at all due to a lack of reasonable protection of privacy. It is the conduct of the individual that removes any need for the judicial oversight of government conduct.

At roughly the same time, the European Court of Human Rights (hereinafter: ECtHR) and the European Court of Justice (hereinafter: ECJ) have repeatedly recognised the need to strengthen judicial oversight of communication traffic data acquisition and storage. In the European context, the requirement for judicial oversight stems from the ECtHR judgement in Malone (*Malone* v. *United Kingdom*, no. 8691/79, 2 August 1984), requiring in essence that traffic data enjoy protection equivalent to the protection against interference with the contents of the communication.

In recent years, in face of perceived terrorist threats, technological marvels and ubiquity of surveillance practices, the ECtHR has not departed from these principles. Recent decisions in Roman Zakharov (*Roman Zakharov* v. *Russia*, no. 47143/06, 4 December 2015 [GC]), and Szabo and Vissy (*Szabo and Vissy*

v. *Hungary*, no. 37137/14, 12 January 2016) testify to that. An illustrative example of its persistence can be found in Szabo and Vissy, reiterating the principles that have guided the court in its past judgements:

> The Court recalls that the rule of law implies, inter alia, that an interference by the executive authorities with an individual's rights should be subject to an effective control which should normally be assured by the judiciary, at least in the last resort, judicial control offering the best guarantees of independence, impartiality and a proper procedure. In a field where abuse is potentially so easy in individual cases and could have such harmful consequences for democratic society as a whole, it is in principle desirable to entrust supervisory control to a judge.

In Szabo and Vissy (§78–79), moreover, the court noted the need for judicial supervision of both, collecting and processing data through surveillance measures, emphasising not only the need for *ex ante* judicial authorisation but also the need for a posteriori judicial (or equivalent) oversight of transferring and sharing of the data.

These are the principles that EU Member States and other signatories to the ECHR must follow. There are, however, instances that appear to rely on the third party doctrine. The Slovenian Constitutional Court, for example, held in February 2014, that no judicial authorisation is required for the police to collect dynamic IP numbers from a seized server and use them to identify the suspect, when the suspect has engaged in sharing forbidden contents through peer-to-peer protocols (Up-540/11, 13 February 2014). The court's reasoning stressed that the applicant effectively waived any protection of privacy in his indiscriminate and mass exchange of digital contents with others, thus demonstrating a lack of any expectation to safeguard his identity.

DNA

The variety of normative solutions when it comes to collecting and processing cellular samples is even more obvious. Some jurisdictions have chosen to submit collecting cellular samples and DNA analysis to judicial oversight. Such a jurisdiction is Germany, where Articles 81a, 81e, 81g and 81f of the German Strafprozessordnung require that bodily examination, genetic examination and DNA analysis (lacking individual's consent) be – in principle – ordered by a judge.

Jurisdictions in Germany's near proximity, within the Council of Europe, however, offer a variety of solutions. Switzerland, for example, uses a mixed system, where it is the police that may order the taking of samples by non-intrusive methods and the creation of DNA profiles from relevant biological materials (Art. 255 [2]) of the Swiss Strafprozessordnung. On the other hand, mass testing (Art. 256 of the Swiss Strafprozessordnung) or taking samples from convicts (Art. 257 of the Swiss Strafprozessordnung) may only be authorised by the court. At the other end of the spectrum is Slovenia, where the taking of samples by non-intrusive means (from suspects or third parties), as well as the decision to perform DNA-analysis is entirely within the competence of the police (Art. 149 of the Slovenian Criminal Procedure Act).

To date, the ECtHR has offered no definitive answer in this respect. The renowned judgement in S. and Marper (*S. and Marper* v. *United Kingdom*, nos. 30562/04 and 30566/04, 4 December 2008 [GC]) has offered an authoritative decision on the issue of indiscriminate retention of fingerprints, cellular samples and DNA profiles of persons suspected but not convicted (§125). It did not find it necessary, however, to decide the issue of the non-existence of independent review put forth by the applicants (§89). Nevertheless, S. and Marper does provide us with a glimpse of safeguards that may be required to guard against risks of abuse and arbitrariness. Its reference to interferences such as telephone tapping, secret surveillance and covert intelligence gatherings is instructive in this respect (§99):

> It is as essential, in this context, as in telephone tapping, secret surveillance and covert intelligence-gathering, to have clear, detailed rules governing the scope and application of measures, as well as minimum safeguards concerning, inter alia, duration, storage, usage, access of third parties, procedures for preserving the integrity and confidentiality of data and procedures for its destruction, thus providing sufficient guarantees against the risk of abuse and arbitrariness.

Some European jurisdictions, at least, find it necessary to subject collecting cellular samples to judicial oversight. We can only speculate, whether the ECtHR, too, would have opted for such a solution. Even so, there is a noticeable contrast between those European jurisdictions that introduced judicial oversight, and the US jurisdiction. In *Maryland* v. *King* (*Maryland* v. *King*, 569 US ___, 2013), the US Supreme Court was asked to rule whether (mandatory) collecting and processing cellular samples of arrestees falls under the Fourth Amendment warrant requirement. As the court acknowledged that using "a buccal swab on inner tissues of a person's cheek in order to obtain DNA samples is a search", it also held that no constitutional requirement for a prior judicial warrant exists, provided that the arrest is valid.

The position of the US Supreme Court marks, therefore, the other side of the spectrum of solutions concerning judicial control over the collection of cellular samples. Interestingly enough, the reluctance (or, the opposite, the perceived need) to provide judicial oversight over collecting cellular samples within a jurisdiction, broadly corresponds to the reluctance (or the need to) provide oversight of collecting communication data. This variety of approaches, as well as the corresponding attitudes towards (a lack of) judicial control of such law enforcement practices, calls for further examination of underlying rationales that led to different requirements in different jurisdictions.

(Why) Is it (not) worth it? The threshold of judicial oversight requirement

Communication traffic data

ECtHR and ECJ

In Malone, the ECtHR was required to answer whether "metering", i.e. using a device to register the numbers dialled on a particular telephone and the time and duration of each call, and submitting the data to the police, amounts to an interference with the right guaranteed under Art. 8 of the ECHR. While the court did not need to decide whether judicial authorisation is required, it did, however, stress that metering records information that is "an integral element in the communications made by telephone".

This reasoning alone may suffice to invoke judicial authorisation in jurisdictions where interference with the privacy of communications is subject to judicial oversight. Such is the case with Slovenia, where Art. 37 (2) of the Slovenian Constitution requires that "only a law may prescribe that on the basis of a court order the protection of the privacy of correspondence and other means of communication and the inviolability of personal privacy be suspended". Thus, relying on Malone, the Slovenian Constitutional Court held in Up-106/05 (2 October 2008):

> It follows from the case law of the European Court of Human Rights … that information on the telephone numbers dialled are considered an integral element of telephone communications.… In view of the above-mentioned, the scope of the protection of communication privacy must be interpreted more broadly such that it also includes information on telephone calls which are an integral element of the communication.… Therefore, obtaining data on the last dialled and last unanswered calls as well as the examination of the content of the short text messages entail an examination of the content and circumstances of the communication and consequently an interference with the right determined in the first paragraph of Article 37 of the Constitution … 10. In accordance with the second paragraph of Article 37 of the Constitution, an interference with the freedom of communication is thus not allowed without a prior court order.

The ECtHR, however, left the question open. In P.G. and J.H. (*P.G. and J.H.* v. *United Kingdom*, no. 44787/89, 29 September 2001), the court, citing Malone, found that metering satisfied the criteria under Art. 8 (2) of the ECtHR without judicial authorisation, distinguishing between metering and interception of communications. In the court's opinion, metering is to be "by its very nature" distinguished from the interception of communications, and the safeguards will depend on the nature and the extent of the interference. The court found no need to require judicial authorisation when the data and their use are strictly limited. In this judgement, the data contained the telephone numbers called from a person's flat between two specific dates.

The impression left by the court in P.G and J.H. is that the communication traffic data is in substance different from data collected by the interception of the communication's content. While this may hold for the specific facts in P.G. and J.H., this differentiation cannot be upheld in cases of mass processing of communication traffic data. The ECJ in Digital Rights (C-293/12 and C-594/12, 8 April 2014) pointed out that processing communication traffic data may very-well have an effect equivalent to the interception of communication's content, that is, an effect on the use of the means of communication and effect on the exercise of the freedom of expression (§28).

What considerations led the ECJ to such a conclusion? The ECJ stressed that the variety of data made available by data retention under the then-valid Directive 2006/24/EC included data on the source of a communication, its destination and location, the date, time, duration, type of a communication and users' communication equipment, the name and address of the subscriber or registered user, the calling telephone number, the number called and an IP address for Internet services, etc. It is the possibility of:

> very precise conclusions to be drawn concerning the private lives of the persons whose data has been retained, such as the habits of everyday life, permanent or temporary places of residence, daily or other movements, the activities carried out, the social relationships of those persons and the social environments frequented by them (§27).

that makes mass communication traffic data retention and processing substantially different from "metering" decisions of the ECtHR and essentially comparable with the interception of the communication's content.

The ECJ decision echoes decisions of Member States' courts, in particular the German Bundesverfassungsgericht 2010 decision (1 BvR 256/08, 1 BvR 263/08 and 1 BvR 586/08, 2 March 2010). The German court also noted that:

> in future with increasing frequency, such storage can make it possible to create meaningful personality profiles and mobility profiles of virtually all citizens. In relation to groups and associations, the data also, in certain circumstances, may make it possible to reveal internal influence structures and decision-making processes.

Two years later, the German court dealt with the constitutionality of attributing telecommunication numbers to their subscribers (1 BvR 1299/05, 24 January 2012), referring to the need for protection of personality, emphasising the need for individuals to appreciate and control government's processing of personal data:

> The guarantee of the fundamental right takes effect in particular when the development of personality is endangered by government authorities using and combining personal information in a manner which persons affected can

> neither fully appreciate nor control.... The extent of protection of the right to informational self-determination is not restricted to information which by its very nature is sensitive and for this reason alone is constitutionally protected. In view of the possibilities of processing and combining, there is no item of personal data which is in itself, that is, regardless of the context of its use, insignificant.... In particular, the protection of informational self-determination also includes personal information on the procedure by which telecommunications services are provided.

It is the evaluation of the impact of traffic communication data storage and access that is the cornerstone of the ECJ's decision in Digital Rights. Its application of the proportionality test, requiring that the data be retained and accessed only when strictly necessary in pursuit of the Directive's objects, finally led the court to clearly outline the requirement of judicial (or equivalent) authorisation:

> The access by the competent national authorities to the data retained is not made dependent on a prior review carried out by a court or by an independent administrative body whose decision seeks to limit access to the data and their use to what is strictly necessary for the purpose of attaining the objective pursued and which intervenes following a reasoned request of those authorities submitted within the framework of procedures of prevention, detection or criminal prosecutions.

The ECtHR followed suite in Szabo and Vissy. It is a case that is to be distinguished from P.G. and J.H.. Here, the court was not faced with a single, limited interference, but noted the possibilities of broad, generalised ("range of persons") and technologically advanced interferences with rights provided by Art. 8 of the ECHR (Szabo and Vissy, §73) and found judicial authorisation (or its equivalent) a fundamental safeguard against abuse and arbitrariness. It is evident that the court was equally concerned with the intrusive nature of interferences in question (§70) and additionally called for an enhancement of safeguards under Art. 8 of the ECHR.

United States

The approach taken by US courts is radically different. Their reliance on the so-called third party doctrine, applying the criteria of reasonable expectations of privacy as developed in *Katz* v. *United States* (389 US 347 (1967)) is conduct-oriented. In the words of the US Supreme Court in *Smith* v. *Maryland*, "it is important to begin by specifying precisely the nature of the state activity that is challenged". The court, similarly to ECtHR in P.G. and J.H., distinguished between acquiring data about dialled phone numbers and acquiring the content of the communications. In applying the Katz test, the court found that 1) no expectation of privacy in general exists as to the numbers dialled, given that subscribers are aware of the practices employed by telephone companies that involve recording and processing records of numbers dialled; and 2) that any expectation of the petitioner could not be deemed reasonable, given that no expectation of privacy is reasonable in respect of information voluntarily turned over to third parties.

Both arguments rest on the fact that an individual willingly submits communications traffic data to telecommunication service providers, third parties. A practical application of the third party doctrine and its focus on the (government) conduct in question can be illustrated by arguments set forth by the government in *ACLU* v. *Clapper* (see Defendant's Memorandum). The government argued the relevance of (1) the conduct of the government agencies; (2) the conduct of the individual subscriber; and (3) the conduct of the service providers.

> The Orders are directed to telecommunications service providers, not to subscribers, and direct the production of what are indisputably the providers' own business records.... Smith is fatal to Plaintiffs' claim that the collection of metadata records of their communications violates the Fourth Amendment. So far as metadata include such information as the times and duration of their calls and the numbers of the parties with whom they spoke, that is information that telephone subscribers voluntarily turned over to their providers. The remaining data, such as trunk identifiers, is information generated by the phone companies themselves.

On the other side, the position of the plaintiff remarkably mirrors the position of the European courts. In their reply to the government, the plaintiffs stress the "content" provided by traffic data and the confidential nature of data aggregated due to long-term recording (see Plaintiffs' Memorandum):

> The kind of surveillance at issue here hands the government a comprehensive record of Americans' associations, revealing a wealth of detail about their familial, political, professional, religious, and intimate relationships – the same kind of information that could traditionally only be obtained by examining the contents of communications.

It is also evident, that such arguments require the US courts to part with the third-party doctrine and look at government practices from a different angle. This poses a rather daunting task, given that the doctrine is well-settled in US case law. To adopt the position of the plaintiffs, the courts would need to re-evaluate the impact of technological advances on safeguards afforded under the Fourth Amendment and to shift from conduct to impact-oriented analysis of government interferences. At this moment, it suffices to say that despite overwhelming (theoretical) criticism (discussed below) and some basis in case law, it is a step yet to be taken by the US courts.

DNA

The glimpsed conduct – impact dichotomy is also reflected in the field of processing DNA data. In general, those legal systems that rely on impact-assessment of collecting and processing cellular samples are more likely to require judicial authorisation. Those that rely on conduct-assessment will more likely find no need for *ex ante* or *ex post* judicial oversight.

German and US jurisdiction, and their contrasting positions on the need to introduce judicial control over processing of DNA data, can be used to illustrate the principle. In *Maryland* v. *King*, the court performed a multi-faceted analysis, demonstrating its perceived need to place an emphasis on either the conduct of government agencies or the conduct of the arrestee. The court thus found that the manner of collecting the inner tissue of a person's cheek by a buccal swab "involves but a light touch on the inside of the cheek", finding that fact "of central relevance" to determining its reasonableness, and proceeded to weigh the privacy related issues with law-enforcement concerns. Equally conduct-focused was the court when responding to claims concerning the processing of DNA samples in the DNA database, CODIS. Central to the court's argument were findings that the processing involves only non-coding DNA that at present do not reveal information beyond identification – nor are the samples tested to any other end. Instructive is the court's understanding that taking DNA samples and their testing falls within the category of cases where the need for a neutral magistrate is minimal, given that limits to such an intrusion are narrowly and specifically defined.

The situation in Germany is quite different. With respect to DNA analysis and storage, the German federal constitutional court, Bundesverfassungsgericht, settled the constitutional benchmarks for DNA processing and collection in 2001. In its 2001 decision (2 BvR 1741/99, 2 BvR 276/00 and 2 BvR 2061/00, 18 January 2001) the court dealt with the constitutionality of safeguards under Art. 81g StPO and found that the requirement of judicial oversight with respect to DNA sampling and analysis is central to the proportionality principle, requiring the courts to ascertain the seriousness of the offence and to take into account the specific circumstances of each case that warrants DNA sampling and analysis. It is a position quite contrary to that taken by the US Supreme Court. Even more, the court's decision applies only to non-coding parts of the DNA; it seems that, for the court, the coding parts of DNA belong to a personal intimate sphere that must remain absolutely free from intrusion (der absolut geschützte Kernbereich der Persönlichkeit; see also Weigend & Ghanayim, 2011, p. 221). In sum, the court – although briefly – based its considerations on the impact of the DNA sampling and analysis for the individual concerned.

Conduct-impact and dignity-liberty: methodological and value dichotomies

An examination of case law approaches to the requirement of the judicial oversight of mass collecting and processing of personal data reveals substantial, methodological and geographical divides in the perception of the legal dimensions to big data processing in the context of criminal justice. The divide is geographical as it reflects two distinct approaches adopted by US and by European courts. In making this distinction I dare not claim the uniformity of approaches in US or in European jurisdictions. The distinction does, however, coincide with markedly different values and methodologies underlying their respective case law.

Nor should I claim that the conduct-impact dichotomy is integral only to mass collecting and processing of personal data. It has been glimpsed and studied before. It has been termed "transatlantic clash" by Whitman (2004), who studied cultural and legal differences in perceptions of privacy across the US and European jurisdiction. He illustrated the distinctions by stressing the US-cherished sanctity of home, free from government's intrusions, and the European-centred concern with individual's public image (name and reputation), noting the right to informational self-determination under German law (Whitman, 2004, pp. 1161–1163). Ross (2007, p. 561 *et seq.*) found a marked distinction between interests at stake when government engages in undercover policing: while German privacy law protects "dignitary interests", the US system focuses on "physical privacy" and "decisional autonomy" (p. 562). Similarly, Jacoby (2006) in her study of technical surveillance measures contrasts the different approaches to privacy protections under US and German jurisdictions: where the US system protects privacy as a negative right to be free from illegal searches and seizures, German law links privacy with human dignity, constructing an "affirmative obligation … to create conditions that foster and uphold the private sphere" (p 491). Likewise, in the context of advanced surveillance technologies (applied in the context of drone surveillance), the conduct-oriented analysis of US courts under the reasonable-expectation-of-privacy test has been contrasted to the dignity-centred approach (likely) to be applicable in the European context (Gorkič, 2012).

To put in perspective the dichotomies (and the divide) that we can glimpse at the comparative level, Solove's typology of privacy conceptions (Solove, 2002a) offers a useful conceptual tool. It exposes the wide range of ideas that have emerged in both theory and judicature. It shows that the issues related to the judicial oversight of processing mass personal data reflect a competition between the concept of privacy-as-secrecy (Solove, 2002a, 1105 *et seq.*) and the concept of personhood.

The third-party doctrine is basically the argument that "once a fact is divulged in public … it can no longer remain private" (Solove, 2002a, p. 1107). As such it is an offspring of the test the US Supreme Court grounded in Katz, the reasonable-expectation-of-privacy (or, should we say, secrecy) and developed, among others, in aforementioned *Smith* v. *Maryland*. *Maryland* v. *King*, too, can be studied in this context. The court's argument that buccal swabs should be seen as a part of a typical post-arrest routine can be restated as a confirmation of complete control over both physical and "informational" aspects of an individual in custody, when no secrets held by the arrestee's body are safe from government's inspection. Hence, the arrestee's "privacy expectations" are necessarily diminished.

On the other side, even if struggling, we find the attempts to frame a different concept of privacy. The concept of personhood, as Solove's analysis of various authors shows, is grounded – or substituted – in terms of "individuality", "autonomy" or "personal dignity" (Solove, 2002a, p. 1116). Solove, citing Freund, shows how such a conception – at least in the US context – has been grounded in Warren and Brandeis's notion of "inviolate personality" (e.g. Warren & Brandeis, 1890, p. 211). Most interestingly, the US Supreme Court, too, has ventured into the concept of privacy-as-personhood, protecting the right of every individual to the possession and control of his own person, free from interference of others "in making certain kinds of important decisions" (Solove, 2002a, citing US case law). Without venturing into the problems of privacy-as-personhood approaches, it is important to note that it closely resembles the concepts that have developed in the European context (see above).

It therefore seems reasonable to conclude that the question of judicial oversight of processing vast amounts of personal data is not merely a matter of (a correct) application of the Katz test. The stakes are much higher, calling for a reconsideration of a general approach to privacy protection (at least in the US context, as perceived from a European perspective). It very much requires a shift from the privacy-as-secrecy to privacy-as-personhood conception, to apply Solove's typology. Solove's (2002a, p. 1151) claim that US Supreme Court's approach to privacy-as-secrecy is ill-suited in the context of *Smith* v. *Maryland*, seems, in particular in the post-Snowden era, to be quite a prophetic one.

The US case law and scholars, however, appear very much to be struggling in determining, if such a shift should indeed take place. Debates on the fate of the third-party doctrine in the pre-Snowden era (e.g. Kerr, 2009; Murphy, 2009) seem particularly out of place, or at least limited to a specific context of law enforcement. The true extent and implications of the third-party doctrine have only become known when the doctrine was applied in the context of national security. Its application in justifying bulk collection and processing of traffic data is symptomatic of the blurring lines between law enforcement and national security interests. At the same time, the court's reluctance to address the constitutional arguments, for example in *ACLU* v. *Clapper*, is a symptom of the court's reluctance to reconceptualise how the Fourth Amendment protects privacy interests (cf. Richards, 2013, p. 1951, commenting on *ACLU* v. *NSA*, 493 F.3d 644, 657 (6th Cir. 2007)). Without a shift in the conceptualisation of privacy, it may very well be impossible for the US courts to assess why surveillance is (can be) harmful (cf. Richards, 2013, p. 1963).

Others appear much more determined. Solove's argument (2002b, p. 1173) that "[i]n order to protect privacy in the Information Age, we must abandon the secrecy paradigm" has been echoed by others. Still, the US judiciary in his analysis in 2002 (Solove, 2002b, p. 1179) appears as divided on the issue as it is (currently) reluctant to act.

This might explain why some appear to have abandoned the Fourth Amendment as the cornerstone of privacy protection against government's intrusions and have focused on arguments originating in the First Amendment and the freedom of expression. Richards' account on the dangers of surveillance (Richards, 2013) is such an example. His shift towards protection of intellectual privacy underlines the political dimension of privacy. He views surveillance practices as an interference with development of new ideas and the freedom of the people to "make up their minds at the ties and place of their own choosing", to freely think, read and privately consult with their confidants (Richards, 2013, p. 1946), all of which are at the core of the freedom of expression. The practices of mass surveillance have reminded us that privacy is, in fact, a political freedom – and is, as such, practised (also) in relation to others.

Solove, too, argued along the same lines (2007, pp. 166–169). In his application of the First Amendment, he found undeniable similarities in (1) obtaining query data on the Internet; (2) obtaining book shop (and, we can add, library) records; (3) obtaining ISP records to identify an anonymous author of a (political) blog post; and (4) obtaining phone call records (grounded in the third party doctrine). What is damaging to the freedoms protected under the First Amendment, is the resulting "chilling effect", i.e. interference with a person's "reading habits and intellectual pursuit" and the possibility of exposing a person's identity to the government.

These arguments all mirror the considerations put forth recently by the European courts. They, too, noted the political dimensions of the right to privacy, underlining the effects of privacy-interferences on the freedom of expression, which goes to the very heart of political freedoms in the European common legal framework (see also Whitman, 2004). More specifically, the connection between the right to privacy and the freedom of expression was first noted in ECJ's Digital Rights judgement. The judgement stressed that indiscriminate retention of communication data may result in persons' "feeling that their lives are subject of constant surveillance" (§37, echoing the opinion of Advocate General Cruz Villalón and in turn, the judgement of the German Bundesverfassungsgericht, 2 March 2010, No 1 BvR 256/08, 1 BvR 263/08 and 1 BvR 586/08). This finding goes hand in hand with ECJ's concern for an effective exercise of the freedom of expression, given a possible impact on the subscriber's use of certain means of communication (§28). These arguments were recently reiterated in ECJ's Tele2 Sverige judgement, No. C–203/15 and C–698/15, 21 December 2016. Even though the importance of the judgement may lie elsewhere, it is crucial to note in this context, the determination of the ECJ to consider communication data retention in the light of fundamental principles governing a democratic, pluralist society (§93).

Arguments relying on the need to protect the freedom of expression tell us that privacy-as secrecy (to borrow Solove's term) is faulty because it forgets that "personhood" of a human being can only be constituted in relation to others. In other words, it is the inviolability of the exchange of ideas and the freedom to develop new ideas – in relation to others – that is in need of protection. Privacy can, in other words, be rephrased in terms of an individual's power to be in control of the relationships and the exchanges he or she chooses to engage in. The "chilling effect", recognised by Solove, Richards and the ECJ, is, in a sense, a deprivation of the individual's power to exercise this control and results in the "normalizing" effects of surveillance practices.

The need and the reluctance of the US courts to approach this very much needed shift in privacy conceptualisation are both evident in recent case law. In Jones *(United States* v. *Jones*, 565 US ___ (2012)), the US Supreme Court had the opportunity to transcend the conceptual boundaries imposed by the Katz privacy-as-secrecy approach. The facts of the case relate to long-term surveillance of a vehicle used by the suspect; thus, employing the Katz approach, it seems impossible to overstep the argument that collecting publicly accessable information on the location of the vehicle should be subject to the Fourth Amendment protections. Indeed, the court's majority did not apply the Katz approach and reverted to applying the Olmstead trespass doctrine (*Olmstead* v. *United States*, 277 US 438 (1928)), focusing on the installation of the tracking device on the suspect's vehicle. Were it to assess the long-term surveillance as a whole, it might recognise the impact of the surveillance and the privacy-related interests in the public sphere. Long-term surveillance and the associated aggregation of data results in a (albeit limited, yet quite comprehensive) "digital biography" (Solove, 2002c) that not only restricts the individual's control of the ways the information is used (Solove, 2002c, p. 1189), but also the individual's ability to freely (intelligently) make decisions, to act and to form relationships with others.

Subsequent case law, however, may offer a glimpse of the changes to come. In Riley (*Riley* v. *California*, 573 US ___ [2014]), the court refused to apply its search-incident-to-arrest exception to a warrant-less search of a cellular phone (and the data it held). It is the arguments that have led the court not to apply the exception that is of outmost interest. They show that the privacy-as-secrecy approach, so embodied in the prevailing Katz approach, can be overstepped by a more rigorous, substantive approach of the privacy-related interest of suspect. The Riley court refused to adopt the arguments set forth in *Maryland* v. *King*, where it found the privacy expectations of arrested suspects are diminished to such an extent that no judicial oversight of collecting and processing DNA samples is required. Instead, it engaged in a substantive analysis of the information stored in today's cellular phones. The following argument is indicative of the court's reasoning:

> First, a cell phone collects in one place many distinct types of information – an address, a note, a prescription, a bank statement, a video – that reveal much more in combination than any isolated record. Second, a cell phone's capacity allows even just one type of information to convey far more than previously possible. The sum of an individual's private life can be reconstructed through a thousand photographs labelled with dates, locations, and descriptions.... Finally, there is an element of pervasiveness that characterises cell phones but not physical records.... An Internet search and browsing history, for example, can be found on an Internet-enabled phone and could reveal an individual's private interests or concerns – perhaps a search for certain symptoms of disease, coupled with frequent visits to WebMD. Data on a cell phone can also reveal where a person has been.

The court's reflections in Riley demonstrates an approach that has transcended the Katz test – and may even make the test obsolete. The privacy interests that the court in Riley recognised are, in fact, interests stemming from interactions with others, whether they be bank records, Internet search and browsing records or person's location – all of the above in possession of a bank, an ISP or exposed to the general public, i.e. to third parties. In this sense, Riley fits very much in Kerr's analysis of the Katz application and his claim that "Katz has only one step", that is, the objective prong (Kerr, 2015). Even though the Riley court seemingly keeps the Katz test alive, its focus rests predominantly on the analysis of the impact a cellular phone search has on an individual. The data held in cell phones are, in fact, a comprehensive "digital biography" (Solove 2002b, p. 1179), revealing the "privacies of life" to the extent that requires the full protection of the Fourth Amendment. Here, subjective expectations and disclosure to third parties gave way to the court's analysis of the societal interests in privacy protection.

DNA-related debates, too, disclose the reluctance of the US courts to recognise the dangers of unsupervised collection of biological samples and the processing. After Riley, it is reasonable to argue, that some secrets of an arrested suspect are, indeed, safe from government's warrantless inspection. And the majority of responses to *Maryland* v. *King* argue, that the court misunderstood what is in fact at stake when buccal swabs are taken and samples analysed (cf. Murphy, 2013).

In contrast to *Maryland* v. *King*, the ECtHR in S. and Marper took a different approach. It held that the mere retention of the cellular material is sufficient to find an interference with the right to respect for private life under Art. 8 of the ECHR (§70). The court also took a more cautious approach towards evaluating the invasiveness of processing cellular materials. It took account of the "rapid pace of developments in the field of genetics and information technology" and the possibilities that private-life interests may be affected in ways that cannot at the time of the decision be anticipated (§71). These considerations lie in stark contrast to *Maryland* v. *King*. The court therefore performed its evaluation of the UK legislation on DNA analysis and the retention of DNA profiles from a broader standpoint: it is the fact that cellular samples are collected and stored that should set the benchmark (§73). In this way, the court would be able to appreciate the full extent of sensitive information that can be extracted by analysing the sample. The court, however, did not find it necessary to do so: unlike the US Supreme Court, it recognised that DNA profiles alone may – and in fact are – used in way that enables familial searching and assessing likely ethnic origin (§75).

The weight of these findings must be viewed in the light of the court's reference to the dangers of stigmatisation of persons who were not convicted of any offence and whose DNA profiles are nevertheless stored (§122). The court found it incompatible with the presumption of innocence when their data is being treated in the same way as the data of convicted persons. This argument should be taken even further: the fact that DNA searches may target – or produce –

information on familial connections or ethnic origin amounts to a suspicion-less searches and racial profiling.

Simoncelli and Krimsky (2007) outlined the risks accompanying the expanding use of DNA databases. They identified several practices that have, seemingly, been overlooked by the US Supreme Court in 2013. First, they warn of the use of DNA dragnets (p. 8 *et seq.*), i.e. mass collection of cellular material from persons that either fit the known profile of the perpetrator or have been placed in the vicinity of the crime scene. Even though their cooperation is seemingly voluntary, their refusal to submit samples may lead the police to treat them as suspects. Such DNA dragnets rest on the assumption that the innocent person will cooperate ("has nothing to hide"), while withholding information is an indicator of guilt. DNA dragnets, therefore, rest on the assumption of guilt and are in stark contrast with the presumption of innocence. They amount to, in fact, a suspicionless search, of which Justice Scalia warned so adamantly in his dissent in King (cf. Roth, 2013, p. 304 *et seq.*).

Second, Simoncelli and Krimsky pointed out that using partial matches while running a DNA profile through a DNA database produces a list of possible relatives of the person whose DNA has been found at the scene of the crime (2007, p. 10; cf. Roth, 2013, p. 307). This is the risk that the ECtHR, too, took into account. The problem of such DNA searches is similar to DNA dragnets. While the issues with DNA dragnets arise at the time of collecting cellular samples, using partial matches creates "suspects" while the data is being processed. Furthermore, using partial matches shows us that the use of cellular material affects a previously undetermined number of persons.

Third, using cellular material may lead to phenotypic DNA profiling (Simoncelli and Krimsky, 2007, p. 11), that is, to making predictions about the physical, behavioural or medical condition of the person. Such profiling goes well beyond the non-coding nature of the STR sequences typically used in DNA profiles. In terms of German case law (see above), it goes beyond the inviolable personable sphere of intimacy.

And finally, Simoncelli and Krimsky (2007, p. 13) warned of covert collection of cellular samples – what Joh termed "abandoned" DNA (Joh, 2006). Collecting "abandoned" DNA goes well beyond collecting biological traces left behind at the scene of the crime. It amounts to a collection of cellular samples of a specific person, suspect or no suspect. Such practices, whether the targeted person is a suspect or not, bear a striking resemblance to other surveillance practices. Of course, it is not the collection of the sample that is hidden from the general public or from the person of interest. The processing and the analysis of the data, however, is. Again, in the light of German case law, it is impossible for an individual to appreciate or to control the information available for the government's inspection.

All of the practices presented by Simoncelli and Krimsky show that the evaluation of taking cellular samples and proceeding with DNA analysis should go well beyond the "buccal swab" and "a light touch in the inside of the cheek". The impact of the DNA sampling and processing is much less physical and much less predictable (than the King court assumed) and much more dependent on the investigation techniques employed by government officials. Even more, the idea that the profiles stored contain only information about non-coding parts of DNA ("junk" DNA), may simply be wrong. If in 2007 Cole and Kaye (Cole, 2007; Kaye, 2007) argued over the predictive values of profiles stored in the CODIS database, Sarkar and Adshead in 2010 reported on research that identified non-coding STR as a marker for red hair (Sarkar and Adshead, 2010, p. 249). Such reports are symptomatic of the so-called function creep, "an incremental enlargement of scope or the addition of new functions... by using the technology or a process in new ways" (Sarkar and Adshead, 2010, p. 249; cf. Joh, 2006, p. 874 *et seq.*).

From the view of an individual whose DNA is collected or analysed, the practices employed and supported by technological advancements simply result in a complete and utter lack of the power to participate or even influence the impact of DNA data processing. Taking into consideration also the profiling potential of "non-coding" DNA, the potential for suspicion-less searches and the covert nature of collecting abandoned DNA, it seems that both, the US and the European jurisdictions, underestimate the effect DNA databases have.

Concluding remarks: the added value of judicial oversight

The different approaches to conceptualising the right to privacy are reflected in the positions within different jurisdictions of the requirement of judicial oversight. The divide in privacy conceptualisations dictates the methodologies courts employ in establishing, whether personal data mass acquisition and processing should be subjected to judicial oversight. What these contrasting positions fail to explain is, what is the added value of judicial oversight?

A typical answer, provided, for example, by the ECtHR in S. and Marper, would refer to the need to curb the risks of arbitrariness. While this may certainly be true, it is necessary to look behind the appearances of a "neutral and detached magistrate" and examine the specific position of individuals subjected to an interference of their right to privacy. The privacy-as-personhood approach reveals the need to see personhood as a relational concept. It is the freedom (and the power) to enter into a relationship with others that gives each individual a claim to "personhood". Inasmuch as the concept of personhood (more so the idea of dignity) may be vague, too narrow or to broad (Solove, 2002a), this dimension can offer an answer as to the added value of judicial oversight. Covert surveillance, bulk collection of communication traffic data and DNA sampling and processing all have a common denominator: the individuals subjected to such practices are simply not offered an opportunity to participate in the decision-making process on how their personal data should be collected and processed. In this sense, they are robbed of the relational dimension of personhood; they are reduced to the proverbial "means to an end".

How can we compensate for the inevitable side effects of such practices? When faced with the need to follow legitimate law-enforcement or national security interests, judicial oversight is the next best alternative, i.e. substituting for the individual's participation with a magistrate's judgement. He, in a sense, is in the best position to speak for the individual, making sure that his privacy interests are recognised and balanced against the interests of society. And the need for this "substitution effect", reflected in the requirement of judicial oversight, grows with the increasingly clandestine nature of surveillance practices, with the exercise of raw power over individual's body and with the increasing alienation and the lack of transparency of personal data processing.

References

Cole, S. A. (2007). Is the " junk" DNA designation bunk? *Northwestern University Law Review Colloquy*, *102*, 54–415.

Defendants' memorandum of law in support of motion to dismiss the complaint. Retrieved from: www.aclu.org/sites/default/files/field_document/govt_motion_to_dismiss.pdf.

Fabbrini, F. (2015). Human rights in the digital age: The European Court of Justice ruling in the data retention case and its lessons for privacy and surveillance in the United States. *Harvard Human Rights Journal*, *28*, 65–95.

Gorkič, P. (2012). The (f)utility of privacy laws: The case of drones? In A. Završnik (Ed.), *Drones and unmanned aerial systems* (pp. 69–92). Heidelberg/New York/Dordrecht/London: Springer.

Jacoby, N. (2006). Redefining the right to be let alone: Privacy rights and the constitutionality of technical surveillance measures in Germany and the United States. *Georgia Journal of International & Comparative Law*, *35*, 433–493.

Joh, E. E. (2006). Reclaiming "abandoned" DNA: The Fourth Amendment and genetic privacy. *Northwestern University Law Review*, *100*, 857–884.

Kaye, D. H. (2007). Please, let's bury the junk: The CODIS loci and the revelation of private information. *Northwestern University Law Review Colloquy*, *102*, 70–81.

Kerr, O. S. (2009). The case for the third-party doctrine. *Michigan Law Review*, *107*, 561–601.

Kerr, O. S. (2015). Katz has only one step: The irrelevance of subjective expectations. *The University of Chicago Law Review*, *82*, 113–134.

Murphy, E. (2009). Case against the case for third-party doctrine: A response to Epstein and Kerr. *The. Berkeley Technology Law Journal*, *24*, 1239–1253.

Murphy, E. (2013). License, registration, cheek swab: DNA testing and the divided court. *Harvard Law Review*, *127*, 161–196.

Plaintiffs' memorandum of law in opposition to defendants' motion to dismiss. Retrieved from: www.aclu.org/sites/default/files/field_document/60._pls._memo_of_law_in_opp._to_defs._mot._to_dismiss_2013.10.01.pdf.

Richards, N. M. (2013). Dangers of surveillance. *The Harvard Law Review*, *126*, 1934–1965.

Ross, J. E. (2007). The place of covert surveillance in democratic societies: A comparative study of the United States and Germany. *The American Journal of Comparative Law*, *55*, 493–579.

Roth, A. L. (2013). Maryland v. King and the wonderful, horrible DNA revolution in law enforcement. *Ohio State Journal of Criminal Law*, *11*, 295–309.

Sarkar, S. P., & Adshead, G. (2010). Whose DNA is it anyway? European Court, junk DNA, and the problem with prediction. *Journal of the American Academy of Psychiatry and the Law Online*, *38*(2), 247–250.

Simoncelli, T., & Krimsky, S. (2007, 9 September). *A new era of DNA collections: At what cost to civil liberties*. Retrieved from: www.acslaw.org/sites/default/files/Simoncelli__Krimsky_-_DNA_Collection__Civil_Liberties.pdf.

Solove, D. J. (2002a). Conceptualizing privacy. *California Law Review*, 90, 1087–1155.

Solove, D. J. (2002b). Access and aggregation: Public records, privacy and the Constitution. *Minnessota Law Review*, *86*, 1137–1209.

Solove, D. J. (2002c). Digital dossiers and the dissipation of Fourth Amendment privacy. *Southern California Law Review*, *75*, 1083–1167.

Solove, D. J. (2007). The First Amendment as criminal procedure. *New York University Law Review*, *82*, 112–176.

Warren, S. D., & Brandeis, L. D. (1890). The right to privacy. *Harvard Law Review*, *4*(5), 193–220.

Weigend, T., & Ghanayim, K. (2011). Human dignity in criminal procedure: A comparative overview of Israeli and German law. *Israel Law Review*, *44*(1–2), 199–228.

Whitman, J. Q. (2004). The two western cultures of privacy: Dignity versus liberty. *Yale Law Journal*, *113*, 1151–1221.

10 Big data and economic cyber espionage

An international law perspective

Maruša T. Veber and Maša Kovič Dine

Introduction

In a globalised and highly competitive economic environment, economic data revealing competitive advantages and disadvantages of states and their companies has become ever so important. Some have even elevated its relevance to the "oil of the 21 century" (The Global Risks Report, 2016, p. 18). States are increasingly becoming aware of the power and value of economic big data and are therefore progressively undertaking different methods to gather massive amounts of secret, publicly unavailable economic information of third states. This information is usually made available to businesses in their own country to aid them in gaining strategic economic advantages on the market. In this respect, cyberspace has become an ultimate tool enabling relatively easy, sophisticated and quick access to large amounts of confidential information, essential for the performance and operation of businesses and economic stability of states.

This chapter focuses on state sponsored theft of massive amounts of economic data from foreign states and companies through cyber means for the benefit of the perpetrating state and assesses the legality of such activities under international law. It argues that economically motivated cyber espionage activities by states and their status under international law should be differentiated from other forms of traditional espionage conducted for military, strategic and security reasons. While the legality of such espionage activities at the international level remains uncertain, we are witnessing important legal developments in the area of economic cyber espionage. Debates related to the legality of traditional espionage are still predominantly limited to academic lubrication, while on the other hand economic cyber espionage has entered the policy arena and received significant attention in inter-state economic relations.

The seemingly limitless scope of economic cyber espionage activities, its inherent transboundary nature, ineffectiveness of national prosecution and non-existence of specific international rules governing the area initiated debates and negotiations among states on finding an international solution to address the issue. States have started to realise that "combating the growing threat of economic espionage and trade secret theft will require" not only "concerted efforts by companies to deter such threats" but also "more robust policy responses and cooperation at the international level" (Calia, Fagan, Veroneau, Vetere, & Eichensehr et al., 2013, p. 6).

After defining economic cyber espionage (Chapter 2) and presenting the pertinence of addressing these activities (Chapter 3) this contribution provides three alternative international legal frameworks in the course of which economic cyber espionage is or could be considered and subsequently also countered. The first one corresponds to current developments at the international level and includes bilateral agreements specifically addressing questions of legality of economic cyber espionage activities (Chapter 4). The second one concerns general international law rules, which are traditionally referred to when determining the legality of espionage activities, with special focus on the principle of non-intervention and a corresponding right of targeted states to resort to countermeasures (Chapter 5). Finally, this contribution considers the possibility of states to address economic cyber espionage through trade policy tools, including World Trade Organisation (WTO) Agreement on Trade-Related Aspects of Intellectual Property Rights (TRIPS Agreement) and free trade agreements (Chapter 6).

Defining economic cyber espionage

No international treaty specifically defines peacetime espionage, cyber espionage or economic cyber espionage.[1] Subsequently there is no commonly agreed authoritative definition of such activities in international law. Espionage in times of war is addressed in Article 46 of the Protocol Additional to the Geneva Conventions of 12 August 1949, and relating to the Protection of Victims of International Armed Conflicts (Protocol I) (1977, 1125 UNTS 3), which determines the status of spies engaging in espionage in times of an international armed conflict (see also Baxter, 1951), but does not provide any definition or further clarification.

For the purpose of this chapter the authors define state sponsored economic cyber espionage as

> transboundary clandestine activity engaged in or facilitated by a state against a computer system or network of another state designed to gain unauthorised access to economic data, such as proprietary information, data collections technology, or other data, for economic advantage.
>
> (adapted from: Canadian Security Intelligence Service, 1994)

This means that economic cyber espionage is a clandestine deceitful intrusion into an information technology system, performed with the purpose of collecting large amount of economic intelligence that includes a wide spectrum of different economic data gathered in a relatively short period of time. This refers to the theft of collections of economic data which relates to policy or commercially relevant economic information including technological data, financial, proprietary commercial and government information. Gathered data may also include e.g. consumer's private information, which can be considered as commercially relevant economic information. While we recognise that the collection of such personal data is controversial from the point of view of international human rights law, this aspect is not the object of this study (for more on this see Fidler, 2015).

The purpose of such state sponsored economic cyber espionage is to use collected data to benefit the economy of a perpetrating state, by gaining comparative advantages for the state's businesses and increasing their competitiveness in the international economic market. Some authors, however, seem to confuse state sponsored economic cyber espionage with industrial espionage (Canadian Security Intelligence service, 1994; Lotrionte, 2015, p. 452; Malawer, 2015, p. 1), which relates to theft of trade secrets between private sector entities, meaning that there is no government involvement. Lack of state involvement makes these espionage activities the subject of private law and will not be further discussed in this chapter.

Inherently, economic cyber espionage is an activity carried out in time of peace perpetrated by a state (or actors whose actions are attributable to the state), whereas the target is another sovereign state and its businesses. Confidential information targeted by economic cyber espionage is stored/available on the cyber infrastructure allocated on the territory of the targeted state, irrespective of whether it is privately or publicly owned. Even though cyberspace seems to be an intangible phenomenon, it is based on infrastructure composed of computer networks and their components, which form the physical architecture of cyberspace and are therefore associated with the territory of a particular state (Delerue, 2016, p. 139). This is derived from the principle of sovereignty in cyberspace as defined in the Tallinn manual on the international law applicable to cyber warfare, which grants each state the right to control cyber infrastructure and cyber activities within its territory and only in certain exceptional cases allows for extraterritorial jurisdiction (Schmitt, 2013a, Rule 1 and 2; see also Heintschel von Heinegg, 2013, 2013, pp. 126, 129; Report of the Group of Governmental Experts, 2013, para. 19–20). What matters for the purposes of this chapter is therefore the location of the servers (infrastructure) storing data targeted by espionage and not the ownership of data as such.

The emergence of cloud computing which enables states and companies to store data on a server that can be located in a third state or anywhere in the world further complicates the question of ownership, sovereignty, jurisdiction and legal interest for the protection of the economic data stored in such clouds. As mentioned above, the physical location of the storage of data is often critical in determining which state has control over and interest in the protection of the data and the prevention of theft (Chertoff, 2011). However, in cases where confidential data of one state is located in the cloud outside its territory, the state will face difficulties in justifying control over such data and taking action against the foreign government that is ordering the theft of them. This issue is subject to much controversy in international law.[2] Further elaboration of this question, however, transcends the scope of this contribution, which focuses on international legal protection of economic big data located on the cyber infrastructure in a territory of a state targeted by economic cyber espionage.

Scope of economic cyber espionage

Since 2012 we are witnessing an increasing rise in economic cyber espionage activities sponsored by states. State reports (Administration Strategy, 2013; Froman, 2015), research reports (Mandiant Report, 2013; Brumfield, 2016), regional documents (Cybersecurity Strategy of the European Union, 2013, p. 3) and leakages about state tools designed for global surveillance such as NSA *Boundless Informant* programmes (Macaskill & Dance, 2013) have confirmed, that state sponsored economic cyber espionage activities are on the rise and increasingly target private and state owned businesses and governments in various sectors. These activities have become so widespread that "the question these days isn't which country commits economic espionage, but which doesn't" (Sepura, 1998, p. 131).[3]

The exact scope of economic cyber espionage activities is extremely difficult to measure, due to specific characteristics of cyberspace: anonymity, difficulty to back trace the perpetrators and fear of states and companies to publicly pronounce that they have been targeted by such activities in order not to lose trust of their citizens/customers. Against this background only rough estimations about the scale and economic losses caused by economic cyber espionage can be made.

While some claim we are witnessing "the greatest transfer of wealth in history" (Rogin, 2012), studies evaluated global loss on account of cyber economic espionage activities to one trillion US dollars per year (McAfee Report, 2013). Recent World Economic Forum Global Risks reports (2015, 2016) place data fraud or theft through technological means amongst top ten risks in terms of likelihood for destabilising the global economy. Furthermore, the European Union documents expressly stipulate that "the increase of economic espionage and state-sponsored activities in cyberspace poses a new category of threats for EU governments and companies" (Cybersecurity Strategy of the European Union, 2013, p. 3). The enlarging scope and threat of economic cyber espionage activities is also recognised in the national documents of states (The IP Commission Report, 2013, pp. 18–22; Cybersecurity strategy for Germany, 2011, p. 3).

Bilateral economic cyber espionage agreements and need for specific regulation of economic cyber espionage at the international level

While it has been argued by some that non-regulation of traditional espionage at the international level contributes to the international peace and is a "functional tool that enables international cooperation" (Baker, 2003, p. 1097) (more on this see below), economic cyber espionage is increasingly undermining the international economic order (Skinner, 2014, p. 1183). Current state practice relating to the conduct of and reaction to economic cyber espionage seems to indicate that such activities are distinct from traditional espionage activities and so is their status under international law.

The problem of economic cyber espionage is especially pertinent in the developed world. According to some reports, numerous private companies in the US, UK, Germany and other countries were victims of intrusions into their information systems and subsequent systematic engagement in theft of confidential collections of big data by third states, especially China (IP Commission Report, 2013, p. 12; Mandiant Report, 2013). The US in particular accuses China of engaging in economic cyber espionage to steal commercial secrets for the benefit of Chinese companies (Kuchler & Sevastopulo, 2014) including penetration into important US financial institutions such as Morgan Stanley and the US Chamber of Commerce (Gorman, 2011). It is not particularly surprising therefore, that the US became an advocator for stronger regulation of these activities at the international level. Since 2014 they have pursued diplomatic, political and legal means to argue for international illegality of state sponsored economic cyber espionage activities.

A major turning point in this respect was the indictment of five Chinese People's Liberation Army hackers on the basis of the national Economic Espionage Act (1996, 18 US Code §1831), which explicitly prohibits stealing and copying, duplicating, downloading, altering, destroying, replicating, transmitting of trade secrets without authorisation for the benefit of a foreign government. In this case hackers were retrieving confidential information from state and non-state actors with the aim of improving China's position in the international economic order. This is the first case in which Chinese officials were publicly indicted for the crime of economic cyber espionage in the US and also the first accusation of state officials of a certain state for espionage activities (Lotrionte, 2015, p. 456). Previous convictions only include cases against US citizens, such as e.g. former Rockwell and Boeing engineer Dongfan Chun (United States Department of Justice, 2009), wherease espionage activities of foreign state officials were mostly considered as unfriendly acts not prohibited under international law (Lafouasse, 2001, p. 127; *Case Concerning Military and Paramilitary Activities In and Against Nicaragua*, 1986, para. 273). Therefore, state practice in terms of reactions to espionage was mostly confined to proclamation of state spies as *persona non grata* (Fidler, 2013a; see below).

Following these events, the US urged "China to halt its persistent theft of trade secrets from corporate computers and engage in a dialogue to establish norms of behaviour in cyberspace" (Nakashima, 2013). The US stressed that such activities are in violation of international law and expressed readiness to use sanctions against these activities in the course of trade and investment relations with China (Nakashima, 2013; see also Administration Strategy on Mitigating the Theft of US Trade Secrets, 2013; Bennett, 2014; Lotrionte, 2015, p. 452).

It has to be mentioned at this point, that the US itself was engaged in highly controversial large scale espionage activities including mass gathering of economic data of foreign companies and foreign financial sectors by the National

Security Agency (NSA) (for more on this see Macaskill & Dance, 2013). Numerous countries raised voices against the legality of the US's mass gathering of data without authorisation of the targeted country (Schmitt & Vihul, 2014, p. 27; Buchan, 2016, pp. 71–72).

An important step to prohibit economic cyber espionage activities at the interstate level was launched in September 2015, when a gentleman's agreement was signed between the US and China on economic cyber espionage. Central to this agreement was the commitment of both states "that neither country's government will conduct or knowingly support cyber-enabled theft of intellectual property, including trade secrets or other confidential business information, with the intent of providing competitive advantages to companies or commercial sectors" (Fact sheet: President Xi Jinping's State Visit to the United States, 2015). Among other things they also agreed to cooperate in investigating and mitigating malicious cyber activities emanating from their territory and to establish a high-level joint dialogue mechanism on fighting cybercrime and related issues. Moreover, both countries stated that they are "committed to making common effort to further identify and promote appropriate norms of state behaviour in cyberspace within the international community" (Fact sheet: President Xi Jinping's State Visit to the United States, 2015). While the agreement is not legally binding and does not create legal effects, it nevertheless represents an important milestone in the economic cyber espionage arena. It is a step forward from the "status quo" of these activities in the past and alludes to a further will of the states to regulate these activities at the international level.

In addition, the US–China talks on economic cyber espionage were followed by similar talks between China and the UK and China and Germany. In the UK–China Joint Statement on Building a Global Comprehensive Strategic Partnership for the 21st Century (2015) both states agreed "not to conduct or support cyber-enabled theft of intellectual property, trade secrets or confidential business information with the intent of providing competitive advantage." Moreover, China and Germany are currently negotiating an agreement to strengthen cybersecurity which includes cyber theft of intellectual property and data security "as both countries seek to upgrade their manufacturing industries with advanced digital technologies" (China, Germany Working on Cybersecurity Deal, 2016). The key element of the agreement will be the obligation to refrain from economic cyber espionage and the establishment of a mechanism for dealing with possible breaches, e.g. when faced with a case of cyber espionage (China, Germany Working on Cybersecurity Deal, 2016).

The willingness of states to regulate this issue has wider dimensions. The G20 Summit in Antalya at the end of 2015 issued a statement in which 20 major world economies acknowledged that internet economy brings both opportunities and challenges to global growth. Furthermore, they made an important statement on economic cyber espionage:

> In the ICT environment, just as elsewhere, states have a special responsibility to promote security, stability, and economic ties with other nations. In

> support of that objective, we affirm that no country should conduct or support ICT-enabled theft of intellectual property, including trade secrets or other confidential business information, with the intent of providing competitive advantages to companies or commercial sectors.
>
> (G20 Leaders' Communiqué Antalya Summit, 2015)

All the above mentioned agreements and statements relate specifically to economic cyber espionage, which clearly indicates a growing perception of states, that economically motivated espionage should be regulated at the international level in the sense of prohibition.

Economic cyber espionage under general international law

In parallel to presented initiatives, which currently have no binding legal value, there is an ongoing debate in the international community concerning which existing international rules could be applicable to economic cyber espionage and what possibilities states have to counter such activities.

This chapter first explains different techniques for the conduct of espionage and substantiates the applicability of general international law rules in cyberspace. It then briefly presents current debates on the legality of traditional espionage under general international law and finally pinpoints the special position of economic cyber espionage in international law, focusing on the principle of non-intervention. We conclude that economic cyber espionage cannot be seen as *per se* lawful and propose the use of countermeasures as a possible response to such activities in cases where a breach of the principle of non-intervention could be established.

Different techniques for conducting espionage

The traditional techniques of intelligence data gathering commonly referred to as human intelligence (HUMINT) are nowadays increasingly being replaced by conducts "from afar", the so called signals intelligence (SIGINT), which commonly use cyberspace for intelligence gathering.

Both types of espionage can be equated in the sense of the objective that is being pursued: acquiring information with an unauthorised intrusion. Specific features of cyberspace, which include enabling transfer of massive quantities of data at low costs and no need for state agents to physically cross state boundaries have stimulated debates about the legality of espionage conducted through cyber means in recent years (see e.g. Ziolkowski, 2013). It is acknowledged that international legal rules apply to both HUMINT and SIGINT techniques of espionage (see below), however, the question of possible different legal implications of conventional and cyber espionage will not be further elaborated in this contribution in so far as special characteristics of cyberspace are not important for determining the legality of economic cyber espionage.

Cyber espionage and applicability of international law in cyberspace

Since 2013 there is an agreement in the international community, that cyberspace is not a *legal lacuna* and indeed the existing rules governing inter-state relations apply in this sphere. This was confirmed by the United Nations Group of Governmental Experts in the Field of Information and Telecommunications in the Context of International Security (Report of the Group of Governmental Experts, 2013), Tallinn Manual (Schmitt, 2013a Part A) and G20 statement (G20 Leaders' Communiqué, 2015). While the fact that international rules apply to cyberspace seems to be undisputed (for critical remarks on argumentative patterns accompanying application of international law to cyberspace see d'Aspremont, 2016), the absence of concrete provisions on the issue of espionage, makes it much more complicated and challenging to determine the legality of economic cyber espionage activities at the international level.

Traditional espionage and its status under general international law

Debates on the (i)legality of espionage in international law are predominately focused on traditional espionage, encompassing espionage activities related to military, strategic and national security, which gained its impetus during the Cold War (see e.g. Buchan, 2015, 2016).

While it is acknowledged that several states prohibit clandestine gathering of data in their national legislation regardless of the means used (see e.g. Espionage Act, 18 US Code, §792), no international treaty explicitly prohibits espionage. There only exist some international rules limiting the scope of espionage, (see Article 19 of the Convention on the Law of the Sea, UN General Assembly, 10 December 1982, 1833 UNTS 3) or those defining treatment of captured foreign spies in times of war (e.g. see Articles 4, 5, 64–76 of the Geneva Convention (IV) Relative to the Protection of Civilian Persons in Time of War, 12 August 1949, 75 UNTS 287). Every state has some sort of intelligence service which is being used for clandestine gathering of military and security related information from foreign states (Lotrionte, 2015, p. 459). Oppenheim, one of the first international legal scholars writing on espionage therefore famously stated that "all states constantly or occasionally send spies abroad, and … it is not considered wrong morally, politically, or legally to do so" (Lauterpacht, 1948, pp. 770, 772). Indeed, countries targeted by espionage in the past did not claim that such activities are contrary to international law and neither did they initiate international legal proceedings with regard to that. In most cases, they merely declared a foreign state official caught when doing espionage activities as *persona non grata* due to their immunities protection (Fidler, 2013a).[4] Against the background of such state practice it is difficult to confirm whether a customary international law rule prohibiting or allowing espionage activities has formed.[5] The status of traditional espionage in international law is therefore unclear (Radsan, 2006, pp. 605–606), which is also reflected in divergent scholarly opinions that can be divided into two groups.

Relying on the *Lotus principle* (SS Lotus case, 1927, p. 19), the first group of commentators perceive espionage as being legal. According to this principle international law leaves states "a wide measure of discretion which is limited only in certain cases by prohibitive rules" whereas in cases where such a rule does not exist "every state remains free to adopt the principles which it regards best and most suitable". Pelican therefore concludes that "cyber espionage, like any other form of espionage, is permissible under international law" (Pelican, 2011, p. 364). Ziolkowski (2013, p. 462) describes the status quo as "on the international level, neither its legality nor its illegality can be established" therefore "states are – in general, and apart from a few specific limitation – free to conduct peacetime espionage activities, by whatever means they choose". It has also been argued that espionage facilitates national security (Chesterman, 2006, p. 1078) and international cooperation, by being a tool by which the state can monitor foreign behaviour (Baker, 2003, p. 1092) and enforce transparency in international relations (Bitton, 2013, pp. 17, 23–25).

On the contrary, the second group of authors argue that espionage activities are in conflict with rules of general international law, in particular the principle of state sovereignty and non-intervention (Wright, 1962; Radsan, 2006, pp. 605–606; von Heinegg, 2013, p. 129: Buchan, 2016, pp. 69–73, 78).

Economic cyber espionage and principle of non-intervention

Regardless of above mentioned debates and the remaining uncertain legal status of traditional, national security related espionage under international law, we argue that economic cyber espionage has certain distinct characteristics and subsequently cannot be perceived as per se lawful (similarly Lotrionte 2015, p. 450). The Tallinn manual concludes on this issue that:

> a State's responsibility for an act of cyber espionage conducted by an organ of the State in cyberspace is not [to] be engaged as a matter of international law unless particular aspects of the espionage violate specific international legal prohibition.
>
> (Schmitt, 2013a, p. 30)

The particular aspect concerning economic cyber espionage analysed in this chapter refers to economically motivated mass gathering of big data collections with the aim of providing strategic and economic advantage of a sponsoring state, which in certain cases could amount to coercive economic interference (similarly Lotrionte, 2015, pp. 492–515). Lawfulness of economic cyber espionage activities is therefore questionable in relation to the general international law principle of non-intervention, which is traditionally understood as prohibiting certain economically motivated interferences in the internal matters of states.[6] It is acknowledged that economic cyber espionage might also be problematic from the point of view of other international legal rules, such as principle of sovereignty and international human rights law, however, this contribution only focuses on the question of the principle of non-intervention since it is considered most relevant for the purpose of this chapter.

The principle of non-intervention, a corollary of the principle of sovereignty, is one of the foundations of the international legal system, allowing states to decide freely on their internal and external affairs, without an external intervention from other states. It is codified in numerous international and regional treaties (Montevideo Convention on Rights and Duties of States, 26. December 1933, 165 LNTS 19; 49 Stat 3097; Charter of the United Nations, Article 2(7), 24 October 1945, 1 UNTS XVI; Declaration on Principles of International Law concerning Friendly Relations and Cooperation among States in accordance with the Charter of the United Nations, 24 October 1970, A/RES/2625(XXV); Conference on Security and Co-operation in Europe, *Final Act of Helsinki*, 1 August 1975; *Draft Declaration on Rights and Duties of States*, UN General Assembly Resolution 375, 6 December 1949, UN Doc. A/RES/375, Article 3; Charter of the Organization of American States, 1948, UNTS 3, Article 19) and is part of customary international law (*Case Concerning Military and Paramilitary Activities In and Against Nicaragua*, 1986, para. 202).

Generally, the principle of non-intervention is twofold and includes the principle of non-interference in the internal or foreign affairs of another state on the one hand and prohibition of the violation of the territory of another state on the other hand (for more on this see Delerue, 2016, p. 160). For the purposes of this chapter the focus is on the former facet of the principle, which was famously defined by the International Court of Justice (ICJ) in the *Nicaragua case* when considering if support of the United States to *contras* in Nicaragua amounted to the violation of the principle of non-intervention:

> [T]he principle forbids all states or groups of states to intervene directly or indirectly in internal or external affairs of other states. A prohibited intervention must accordingly be one *bearing on matters in which each state is permitted, by the principle of state sovereignty, to decide freely*. One of these is the choice of a political, economic, social and cultural system, and the formulation of foreign policy. *Intervention is wrongful when it uses methods of coercion* in regard to such choices, which must remain free ones.
> (*Nicaragua case*, 1986, para. 205 [emphasis is authors])

On the basis of this judgement three elements of the principle of non-intervention are generally discerned: first, the act must be committed by a state or attributable to it, second the act means an intervention into another state's sovereign affairs and third, this act is coercive.

The first element relates to the entangled question of attribution in international law, which is not the object of the present study (see Articles on Responsibility of States for Internationally Wrongful Acts [ARSIWA] 2001, Article 2 and Chapter II and Chapter 4.4). In relation to the second element we have already determined in the beginning that any information stored/available on the cyber infrastructure allocated on the territory of the targeted state, irrespective of whether it is privately or publicly owned will be subject to state protection and its sovereign governance (Schmitt, 2013a, Rule 1 and 2). As regards the third element, coercion refers to action that "is taken by one state to secure a change in the policies of another" (Jamnejad & Wood, 2009, pp. 347–348). Such coercion may also be economic (Bowett, 1972; Lillich, 1975), which is corroborated by some UN documents. *Declaration on Principles of International Law concerning Friendly Relations and Cooperation among States in accordance with the Charter of the United Nations* (1970) stipulates that "no State may use or encourage the use of economic political or any other type of measures to coerce another State in order ... to secure from it advantages of any kind" (see also *Declaration on the Inadmissibility of Intervention in the Domestic Affairs of States and the Protection of Their Independence and Sovereignty*, 1965; *Definition of Aggression*, 1974). When considering what level of economic coercion is necessary to breach the principle of non-intervention, Jamnejad and Wood (2009, p. 370) rightfully conclude that it is the impact of coercive measures on the target state that define whether a breach has occurred (see also McDougal and Feliciano, 1958, p. 794). While indeed mere collection of data may not be seen as *per se* coercive in terms of violating the principle of non-intervention, data collected through means of espionage, may be used to coerce another state and affect its freedom of decision thereby perpetrating unlawful intervention (Delerue, 2016, p. 171). In cases of economic cyber espionage large amounts of confidential economic data is gathered systematically and later used (disclosed) by the perpetrating states to advance its economy. This has an inevitable effect on target state's policies, its financial stability and its position on global markets. Having its confidential data reviled, a state will have to adopt further decisions regarding its policies and measures which otherwise would not have been adopted. As Lotrionte (2014, p. 473) puts it:

> Economic espionage is about economic competition, the goal being to prevent the target from advancing economically. It is not about collecting information an adversary tried to keep secret in order to inform the making of policy; rather, it is about stealing property and information to provide domestic companies with an economic advantage, disadvantaging foreign companies, and eviscerating any competition.

Buchan (2016, p. 77, footnote 69; see also Watts, 2015, p. 256) similarly argues that "where information obtained as a result of cyber espionage is subsequently used to exert influence over the victim state, a violation of the non-intervention is likely to occur". However, the final assessment depends on the circumstances of a particular case, scale and nature of economic data that has been gathered and use of such data by the perpetrating state. Against this background authors believe that economic cyber espionage cannot be perceived as *per se* lawful (on the contrary see Fidler, 2013a).

The use of countermeasures

Recognising that economic cyber espionage activities may amount to a violation of the principle of non-intervention is important, as it gives the victim state a recourse for action, regardless whether international law currently regulates economic cyber espionage. On the basis of the Law of State responsibility, states are namely responsible for breaches of international law (ARSIWA) and in such instances, the affected State may take recourse to proportionate countermeasures (ARSIWA, Article 22 and Chapter II). They provide for a means of "self-help" enabling states to react to international wrongful acts of other states. An important characteristics of countermeasures relates to the fact that they may take a variety of forms and need not be reciprocal to prior wrongful cyber espionage activities.

Countermeasures have been recognised by the ICJ and other international tribunals and are considered as part of customary international law (*Naulilaa Incident Arbitration (Portugal* v. *Germany)*, 1928; *Gabčíkovo-Nagymaros Project (Hungary v. Slovakia)*, 1997, para. 82–83; *Nicaragua case*, 1986 para. 249; *Air Services agreement of 27 March 1946 between the United States of America and France*, 1978, pp. 443–446). The purpose of countermeasures is to return a situation to lawfulness (ARSIWA Article 49(1)). They are therefore of a temporal nature and "must be as far as possible reversible in their effects in terms of future legal relations between the two States" (ARSIWA with commentaries, 2001, p. 283; ARSIWA, Article 52 and 53).

For a countermeasure to be lawful it has to be a response to an international wrongful act consisting of a breach of an intentional obligation which has to be attributed to a particular state (ARSIWA, Article 2, 4–11; see also Schmitt, 2013a, Rule 5, 6 and 7; *Nicaragua case*, 1986, para. 115). Moreover, international law places further strict restrictions on the use of countermeasures, such as procedural obligations (ARSIWA Article 43 and 52 [1]b; *Gabčíkovo-Nagymaros Project*, 1997, para. 84);[7] proportionality (ARSIWA Article 51; *Gabčíkovo-Nagymaros Project*, 1997, para. 85; *Air Services agreement*, 1978, para. 83); prohibition of the use of force (ARISWA Article 50 [1]; *Corfu Channel case*, 1949, p. 35; *Declaration on Principles of International Law concerning Friendly Relations*, 1970, para. 6.); respect for obligations relating to protection of fundamental human rights (ARSIWA Article 50 [1]b); and non-violation of peremptory norms of international law (ARSIWA Article 50 [1]).

In cases of economic cyber espionage, a prior breach would relate to the violation of the principle of non-intervention. However, the applicability of countermeasures to cases of economic cyber espionage is limited. It can be extremely challenging to prove that a certain breach of the principle of non-intervention through the conduct of economic cyber espionage is attributable to another sovereign state (ARSIWA, Article 2 and Chapter II) and can thus trigger a response through the adoption of countermeasures. This is especially true because groups of individuals may use different techniques to create the impression that a state was behind certain cyber activities (e.g. "spoofing") (Hinkle, 2011; Calia et al., 2013,

p. 4; Schmitt, 2013a, p. 668; Schmitt, 2014). With regard to that, it was already recognised by the Tallinn Manual that "the mere fact that a cyber operation has been launched or otherwise originates from governmental cyber infrastructure is not sufficient evidence for attributing operation to that State", it is instead a mere indication (Schmitt, 2013a, p. Rule 7). Schmitt (2013b, p. 660) therefore rightly concludes that due to various limitations accompanying the use of countermeasures they cannot be considered as "panacea" for inter-state issues arising in cyberspace. Moreover, recourse to self-help may trigger escalation of coercive measures between states further destabilising their relations instead of lessening the tensions.

Use of trade policy tools

Apart from the possibility of states to regulate economic cyber espionage in bilateral agreements and to use decentralised reactions in the form of countermeasures, existing (or future) trade policy tools could also be used to address economic cyber espionage. The US official documents expressly mention this option: "The Administration will utilize trade policy tools to increase international enforcement against trade secret theft to minimize unfair competition against US companies" (Administration Strategy, 2013, p. 4).

Two possibilities are being mentioned in this part of the contribution: first option presupposes relying on the World Trade Organisation (WTO) system and its agreements, which include provisions on national treatment and protection of trade secrets that might be relevant for economic cyber espionage, whereas the second option includes provisions on economic cyber espionage in free trade agreements that are currently being negotiated.

Multilateral World Trade Organisation system

Existing trade agreements within the WTO could offer some answers to the legality of economic cyber espionage. Agreement on Trade-Related Aspects of Intellectual Property Rights (TRIPS Agreement) provides the minimum standards on the protection of intellectual property rights (Taubman, Wager, & Watal, 2012, p. 10) including trade secrets and undisclosed information. One of the advantages of addressing economic cyber espionage through WTO is the fact that it has a broad membership and a well-developed and effective dispute settlement system (TRIPS Article 64; Merrills, 2011, p. 233). The TRIPS agreement was adopted in 1995 and it is not surprising that it does not have specific provisions on economic cyber espionage activities, but it does include provisions that could be applicable to such activities. Gathering of economic big data could be in contradiction with Article 3 on national treatment and Article 39 providing for minimum standards of protection of undisclosed information.

Some countries already indicated the possibility of pursuing a case in front of the WTO dispute settlement system to mitigate state-sponsored stealing of economic data. The US policy in this respect is clear. While they stress that "the most efficient and preferred manner of resolving concerns is through bilateral dialogue" (Froman, 2015, p. 4), it is expressly stated in the US 2013 special 301 Report that "where these efforts are unsuccessful, the United States will not hesitate to use the WTO dispute settlement procedures, as appropriate" (Froman, 2015, p. 4; see also Malawer, 2015, p. 3). In the past, the US already initiated proceeding against China on the basis of the TRIPS agreement in the case of *China – Measures Affecting the Protection and Enforcement of Intellectual Property Rights* (2009).

Scholarly opinion on this issue is, however, divided. Some believe that litigation through the WTO dispute settlement system on the basis of the TRIPS agreement could provide for one of the "most promising and immediate remedy" (Malawer, 2015) or see the WTO as the "most effective institution" (Skinner, 2014, p. 1165) for addressing these issues. Even though there is no express economic espionage prohibition in the TRIPS agreement some scholars argue that states could make use of non-violation complaint since "the letter and spirit of the agreements indicate that theft of trade secrets are prohibited" and "such theft undermines the purpose of these agreements – to create a fair trade regime among member states" (Skinner, 2014, p. 1197; Lotrionte, 2015, p. 527). It has to be stressed, however, that advocating for the possibility of non-violation complaint on the basis of Article XXIII GATT in the course of which member states can start proceedings whether or not the contemptuous measure conflicts with the provisions of WTO Agreements is legally unfounded. There exists a "moratorium" in conjunction to non-violation complaints and TRIPS agreement, meaning that there is an agreement among states not to use TRIPS non-violation and situation complaints (Taubman, Wager, & Watal, 2012, pp. 159–160). On the other side of the spectrum, scholars consider finding a solution to economic cyber espionage through WTO proceedings "not convincing legally or politically" (Fidler, 2013b) mostly on account of the territorial nature of intellectual property rights protection.

While indeed intellectual property rights are essentially territorial, the viability of this principle of territoriality is gradually put into question by increased trans-border activities through cyberspace (Rahmatian, 2015, p. 79). The strict territorial nature of intellectual property rights is also diminished by their increased international legal regulation. Even though intellectual property is a form of private property and is essentially governed by private law of sovereign states, international law (e.g. TRIPS agreement) introduces minimum standards for their protection, putting constraints on individual states (Rahmatian, 2015, pp. 72, 78). Economic cyber espionage could therefore be addressed through the WTO system indirectly, through states' obligations to ensure protection of intellectual property rights on the basis of Article 3 or 39 of the TRIPS agreement.

Article 3: National Treatment Clause

The general provisions and basic principles in Part I of the TRIPS agreement include Article 3, entitled national treatment, according to which "each Member shall accord to the nationals of other Members treatment no less favourable than that it accords to its own nationals with regard to the protection of intellectual property". This article makes sure that member states do not discriminate between domestic and foreign companies when it comes to intellectual property rights. It is clearly applicable to situations in which state authorities steal data from foreign companies, whereas their own companies would remain protected from such stealing of secrets (see also Fidler, 2013b). However, it remains questionable whether this applies to state sponsored stealing of big data collections by foreign states and companies located outside the territory of a perpetrating state (Fidler, 2013b). The main argument against the application of this provision to trans-boundary state sponsored economic cyber espionage is the fact, that intellectual property rights are traditionally territorially protected. To avoid the question of extraterritorial protection of intellectual property rights, it may be argued, that discrimination in such cases happens on the level of dissemination. Following this argument discrimination happens because data gathered extraterritorially are then provided only to selected state companies, excluding foreign companies from having this information. In this case foreign companies within the perpetrating state have less favourable treatment compared with state-owned companies. Effects of extraterritorial stealing of data are seen on the territory of a perpetrating state, therefore extraterritorial gathering of data is not a problem anymore (similarly Malawer, 2015, p. 5). It would, however, be very challenging to prove that this discrimination is happening on a systematic level, since this is not an official policy or measure of member states, but is rather done in a clandestine way (see also Malawer 2015, p. 4; Strawbridge, 2016, pp. 839–845).

Article 39: Protection of undisclosed information

Part II of the TRIPS agreement sets out minimum standards of protection of intellectual property, which includes protection of undisclosed information in Article 39 and is based on Article 10bis of the Paris convention (1967) regulating unfair competition (for more on this see Strawbridge, 2016, p. 847). According to Article 39, member states are to protect undisclosed information by way of ensuring effective protection against unfair competition. This article gives all natural and legal persons the right to "have the possibility of preventing information lawfully within their control from being disclosed to, acquired by, or used by others without their consent in a manner contrary to honest commercial practices". Information protected by this provision has to be secret; it has to have commercial value because it is secret and is subject to reasonable steps under the circumstances to keep it secret (TRIPS Article 39, para. 2). Article 39 further defines practices that can be understood as being contrary to honest commercial practices: "breach of contract, breach of confidence and inducement to breach, and includes the acquisition of undisclosed information by third parties who knew, or were grossly negligent in failing to know, that such practices were involved in the acquisition". Looking at this provision, espionage as was defined at the beginning of this chapter seems to fit into the category of practice that is contrary to honest commercial practices and means at least a breach of confidence between states. During TRIPS negotiations there was even a consensus that espionage is a practice that inherently constitutes a manner contrary to honest commercial practices (Peter & Michaelis, 2009, pp. 643–644).

In accordance with Part III of the TRIPS agreement member states have to introduce enforcement procedures enabling prompt and effective action against infringements of intellectual property rights under the TRIPS agreement (Pauwelyn, 2010, pp. 413, 421). Intellectual property rights, including collections of big data, are of such nature, that their owner is "entitled to prevent others from undertaking certain acts without his or her authorisation" (Taubman et al., 2012, p. 136). The owners must have the possibility to "stop infringement and prevent further infringement, as well as to recover the losses incurred from infringement" (Taubman et al., 2012, p. 137 and Chapter VII; TRIPS, Articles 41–61).

Indeed, the WTO obligation to afford a "possibility" of preventing information from being disclosed to without consent refers to "WTO members behaviour within its territory towards national of other WTO members doing business in that territory" (Fidler, 2013b). But there may be a possibility of bringing the case before the WTO, indirectly addressing state sponsored cyber economic espionage activities on the account of ineffective protection and enforcement of trade secrets in a particular country.[8] Regarding Chinese economic cyber espionage, US 2013 special 301 Report, for example, recognises that obtaining enforcement of intellectual property rights in China is very challenging and has been made worse with the possibility of cyber theft coming from actors located on Chinese territory (Froman, 2015, pp. 31–32). It is stressed that "available remedies under Chinese law are difficult to obtain, given that civil, administrative, and criminal enforcement of trade secrets theft remains severely constrained (Froman, 2015, p. 32)." The report refers to both steeling of data by China within Chinese territory as well as allegations of stealing economic data in third States and urges that the "Chinese Government take serious steps to put an end to these activities and to deter further activity by rigorously investigating and prosecuting thefts of trade secrets by both cyber and conventional means" (Froman, 2015, pp. 32, 34). This indicates the recognition of China's obligation to protect from big data theft on its territory and the possibility of applying Article 39. in economic cyber espionage cases.

However, it remains questionable how the WTO Dispute settlement body would deal with this issue and whether it would interpret the "possibility" in a way of the existence of legal proceedings or their effectiveness. There yet exists no relevant jurisprudence on the question of undisclosed information under Article 39[9] to give a definite answer on this issue.

Free trade agreements

Apart from the existing TRIPS provisions that could provide a way for affected states to counter economic espionage activities, there is a growing trend in the field of free trade agreements to include provisions related to trade secrets and hence big data theft. If this trend continues, it could provide a way towards new global standards enabling states to advance protection of data with economic value and offer them effective dispute settlement resolution.

Many countries are currently negotiating major trade agreements such as the Trans-Pacific Partnership Agreement (TPP) and Trans-Atlantic Trade and Investment Partnership (TTIP), which open an opportunity for the advancement of protection of economically sensitive data and possibly creation of uniform standards regarding economic cyber espionage (Merkin, 2013). TPP, for example, includes Chapter 18 on Intellectual Property, which expressly addresses trade secrets theft through cyberspace:

> The Intellectual Property chapter requires TPP Parties to provide for the legal means to prevent misappropriation of trade secrets, including misappropriation conducted by State-owned enterprises. It also requires TPP Parties to establish criminal procedures and penalties for trade secret theft, including by means of cyber-theft.
>
> (Trans Pacific Partnership)

As regards the current negotiations on TTIP, the European Commission is of the view that trade secrets will have a heightened level of relevance in the TTIP negotiations, especially due to the recent NSA activities in Europe. The EU and US definitely have a common interest in pursuing protection of trade secrets against other states and also among themselves (Protection against the unlawful acquisition of undisclosed know-how and business information, 2013).

These passages show, that states are increasingly addressing economic cyber espionage activities in free trade agreements and there is a growing perception on the illegality of such activities. Future disputes arising from economic cyber espionage activities could be addressed through means of dispute settlement as provided in these agreements.

Conclusion

Big data theft, sponsored by governments with the intention to gain secret information to aid its companies and give them a competitive advantage over foreign companies on the international markets, is becoming a common occurrence in the international economic arena. An unprecedented increase of state practice in recent years both in terms of perpetration of such economic cyber espionage activities, as well as concerning condemnation of such activities by states, indicates the belief of states that such practice is illegal under international law. There are no international legally binding documents that would confirm this belief, however, steps have been taken towards rectifying this problem, both in terms of development of new rules, as well as in applying existing international law rules to economic cyber espionage and big data theft.

An example of the possible emergence of new rules addressing the issue is the recently signed gentleman's agreement on economic cyber espionage between the US and China, and current negotiations between China and some European countries (Germany and the UK) on bilateral agreements prohibiting state sponsored economic espionage activities.

This chapter also considered the legality of state sponsored economic cyber espionage under general international law. While international legal scholarship is increasingly divided on those arguing for legality and others claiming illegality of traditional espionage activities, this contribution concludes that economic cyber espionage differs from traditional espionage for e.g. national security purposes and will in certain instances amount to economic coercion prohibited by the principle of non-intervention. Subsequently this contribution points out the possibility of the use of countermeasures to counter economic cyber espionage activities.

Another possibility for addressing the issue of economic cyber espionage is through trade policy tools, such as the TRIPS agreement and free-trade agreements. While the likelihood of having proceedings regarding this economic cyber espionage before the WTO seems to be small, states are increasingly considering this issue in the course of free trade negotiations. This may lead to the creation of uniform international standards regarding mass gathering of economic data by states.

It can be concluded that while there is reticence in the international community regarding lawfulness of espionage as such, this cannot be said for economic cyber espionage. States react differently and are much more proactive when it comes to economic cyber espionage activities and seem to perceive it as a question different from other types of espionage. Referring to the general principles of international law and current activities of states in the course of bilateral agreements and trade agreements, we conclude, that there is a growing perception amongst states on the need for regulating state sponsored economic cyber espionage at the international level.

Notes

1 The only exception is espionage in times of war which is addressed in Article 46 of the Protocol Additional to the Geneva Conventions of 12 August 1949, and relating to the Protection of Victims of International Armed Conflicts (Protocol I), 8 June 1977, 1125 UNTS 3, which determines the status of spies engaging in espionage in times of International Armed Conflict. For wartime economic espionage see also Baxter, 1951.

2 This was partly addressed in the recent *East Timor* v. *Australia case* (2014, para. 52), where Australia gathered information located in the territory of Australia but owed by the legal counsel of East Timor. The court granted a provisional order requiring "Australia not to interfere in any way in communications between Timor-Leste and its legal advisers". Buchan (2016, p. 76) makes an analogy to data stored in the clouds and concludes that states storing such data "have the right to have that information protected", however, he excludes information having commercial purposes.

3 Citing counter-intelligent agent George Lepine as cited by Ian McGugan, *The Spy Who Came in for the Gold*, Can. Bus., 1 May 1995, p. 99.

4 However, spies have been receiving different treatment if caught in times of war. See Note 1.

5 Arguing for permissive rule of customary international law: Baker, 2003, pp. 1094–1095. Arguing against the existence of custom permitting espionage: Buchan, 2016, pp. 81–85.
6 Tallinn Manual 2.0, which is to be published in 2017, is expected to deal with peacetime espionage and it remains to be seen how it will qualify economic cyber espionage activities under international law.
7 The procedural obligation of exhaustion of peaceful means of dispute settlement before resorting to countermeasures is not yet established in the customary international law. *Air Services agreement*, 1978, para. 91.
8 To raise the protection of trade secrets in European countries, the EU made a proposal for a Directive on Trade Secrets, which is based on Article 39 of TRIPS and focuses on illegal access and illegal interception of electronic data. Proposal for a Directive of the European Parliament and of the Council on the protection of undisclosed know-how and business information (trade secrets) against their unlawful acquisition, use and disclosure /* COM/2013/0813 final – 2013/0402 (COD).
9 Until now, there was only one proceeding initiated before the WTO on the basis of Article 39, but it was based on its third paragraph which is not relevant for our research. *Argentina – Patent Protection for Pharmaceuticals and Test Data Protection for Agricultural Chemicals*, 1999; *Argentina – Certain Measures on the Protection of Patents and Test Data*, 2000.

References

Books

Lauterpacht, H. (Ed.). (1948). *Oppenheim's international law* (7th edn, Vol. I). London: Longman's, Green & Co.

Merrills, J. G. (2011). *International dispute settlement*. Cambridge: Cambridge University Press.

Osula, A.-M., & Rõigas, H. (Eds.). (2016). *International cyber norms: Legal, policy & industry perspectives*. Tallinn: NATO CCD COE Publications.

Schmitt, M. N. (2013a). *Tallinn manual on the international law applicable to cyber warfare*. Cambridge: Cambridge University Press.

Taubman, A., Wager, H., & Watal, J. (2012). *A handbook on the WTO TRIPS agreement*. Cambridge: Cambridge University Press.

Tsagourias, N., & Buchan, R. (Eds.) (2015). *Research handbook on international law and cyberspace*. Cheltenham: Edward Elgar Publishing.

Chapters in books

Buchan, R. (2015). Cyber espionage and international law. In N. Tsagourias, & R. Buchan (Eds.), *Research handbook on international law and cyberspace* (pp. 168–189). Cheltenham: Edward Elgar Publishing.

Buchan, R. (2016). The international legal regulation of state-sponsored cyber espionage. In A.-M. Osula, & H. Rõigas, (Eds.), *International cyber norms: Legal, policy & industry perspectives* (pp. 65–86). Tallinn: NATO CCD COE Publications.

Fidler, D. P. (2015). Cyberspace and human rights. In A.-M. Osula, & H. Rõigas, (Eds.), *International cyber norms: Legal, policy & industry perspectives* (pp. 94–117). Tallinn: NATO CCD COE Publications.

Peter, M., & Michaelis, M. (2009). Section 7: Protection of undisclosed information. In P.-T. Stoll, J. Busche, & K. Arend (Eds.), *WTO – Trade-related aspects of intellectual property rights* (pp. 631–647). Max Planck Commentaries on World Trade Law, Vol. 7. Leiden: Martinus Njihoff Publishers.

Rahmatian, A. (2015). Cyberspace and intellectual property rights. In N. Tsagourias, & R. Buchan (Eds.), *Research handbook on international law and cyberspace* (pp. 72–93). Cheltenham: Edward Elgar Publishing.

Schmitt, M. N. (2013b). Cyber activities and the law of countermeasures. In K. Ziolkowski (Ed.), *Peacetime regime for state activities in cyberspace* (pp. 659–688). Tallinn: International Law, International Relations and Diplomacy, NATO CCD COE Publication.

Watts, S. (2015). Low-intensity cyber operations and the principle of non-intervention. In J. D. Ohlin, K. Govern, & C. Finkelstein (Eds.), *Cyber war: Law and ethics for virtual conflicts* (pp. 249–271). Oxford: Oxford University Press.

Ziolkowski, K. (2013). Peacetime cyber espionage – new tendencies in public international law. In K. Ziolkowski (Ed.), *Peacetime regime for state activities in cyberspace* (pp. 659–690). Tallinn: International Law, International Relations and Diplomacy, NATO CCD COE Publication.

Articles

Baker, C. D. (2003). Tolerance of international espionage: A functional approach. *Am. U. Int'l L. Rev., 19*, 1091.

Baxter, R. R. (1951). So-called unprivileged belligerency: Spies, guerrillas, and saboteurs. *Brit. YB Int'l L., 28*, 323.

Bitton, R. (2013). Legitimacy of spying among nations, The. *Am. U. Int'l L. Rev., 29*, 1009.

Bowett, D. W. (1972). Economic coercion and reprisals by states. *Va. J. Int'l L., 13*, 1.

Chesterman, S. (2006). The spy who came in from the Cold War: Intelligence and international law. *The Michigan Journal of International Law, 27*, 1071–1130.

d'Aspremont, J. (2016). Cyber operations and international law: An interventionist legal thought. *J Conflict Security Law, 21*(3), 575–593.

Fraumann, E. (1997). Economic espionage: Security missions redefined. *Public Administration Review*, 303–308.

Heintschel von Heinegg, W. (2013). Territorial sovereignty and neutrality in cyberspace. *International Law Studies, 89*, 123–156.

Hinkle, K. C. (2011). Countermeasures in the cyber context: One more thing to worry about. *Yale Journal of International Law, 37*, 11–21.

Irion, K. (2012). Government cloud computing and national data sovereignty. *Policy & Internet, 4*(3–4), 40–71.

Jamnejad, M., & Wood, M. (2009). The principle of non-intervention. *Leiden Journal of International Law, 22*(02), 345–381.

Lafouasse, F. (2001). L'espionnage en droit international. *Annuaire français de droit international, 47*, 63–136.

Lillich, R. B. (1975). Economic coercion and the international legal order. *International Affairs (Royal Institute of International Affairs 1944–), 51*(3), 358–371.

Lotrionte, C. (2015). Countering state-sponsored cyber economic espionage under international law. *NCJ Int'l L. & Com. Reg., 40*, 443–541.

Malawer, S. S. (2015). Chinese economic cyber espionage: US litigation in the WTO and other diplomatic remedies. *Georgetown Journal of International Affairs*, International engagement on Cyber V. Retrieved from www.globaltraderelations.net/images/

Malawer.China_Cyber_Economic_Espionage_Lead_Article_Georgetown_Int_l_Affairs_J._June_2015_.pdf.

McDougal, M. S., & Feliciano, F. P. (1958). International coercion and world public order: The general principles of the law of war. *The Yale Law Journal*, *67*(5), 771–845.

Macaskill, E., & Dance, G. (2013, 1 November). NSA files decoded. What the revelations mean for you. *Guardian*. Retrieved from: www.theguardian.com/world/interactive/2013/nov/01/snowden-nsa-files-surveillance-revelations-decoded#section/1.

Pauwelyn, J. (2010). The dog that barked but didn't bite: 15 years of intellectual property disputes at the WTO. *Journal of International Dispute Settlement*, idq001.

Pelican, L. (2011). Peacetime cyber-espionage: A dangerous but necessary game. *CommLaw Conspectus*, *20*, 363–386.

Radsan, A. J. (2006). Unresolved equation of espionage and international law, *The Michigan Journal of International Law*, *28*, 595, 605–606.

Saias, M. A. (2014). Unlawful acquisition of trade secrets by cyber theft: Between the proposed directive on trade secrets and the directive on cyber attacks. *Journal of Intellectual Property Law & Practice*, *9*(9).

Schmitt, M. N. (2014). Below the threshold cyber operations: The countermeasures response option and international law. *Virginia Journal of International Law*, *54*.

Schmitt, M. N., & Vihul, L. (2014). The nature of international law cyber norms. Tallinn Paper No. 5, NATO CCDCOE.

Sepura, K. (1998). Economic espionage: The front line of a new world economic war. *Syracuse J. Int'l L. & Com.*, *26*, 127–150.

Skinner, C. P. (2014). An international law response to economic cyber espionage. *Connecticut Law Review*, *46*, 1165–1207.

Strawbridge, J. (2016). The big bluff: Obama, cyber economic espionage, and the threat of WTO litigation. *Georgetown Journal of International Law*, *47*, 833–865.

Wright, Q. (1962). Espionage and the doctrine of non-intervention in internal affairs. *Essays on Espionage and International Law*, 793–798.

Other

Administration strategy on mitigating the theft of US trade secrets (2013). White House. Retrieved from www.whitehouse.gov/sites/default/files/omb/IPEC/admin_strategy_on_mitigating_the_theft_of_u.s._trade_secrets.pdf.

Articles on Responsibility of States for Internationally Wrongful Acts (ARSIWA) with commentaries (2001). *Yearbook of the International Law Commission*, 2001, vol. II (Part Two).

Bennett, C. (2014, 11 October). Obama urges China to stop cyber theft. *The Hill*. Retrieved from http://thehill.com/policy/cybersecurity/223555-obama-urges-china-to-stop-cyber-theft.

Brumfield, J. (2016). *Data breach investigations report*. Retrieved from http://news.verizonenterprise.com/2016/04/2016-data-breach-report-info/.

Calia, K., Fagan, D., Veroneau, J., Vetere G., Eichensehr, K., Cilluffo, F., & Beckner, C. (2013, September). *Economic espionage and trade secret theft: An overview of the legal landscape and policy responses*. Retrieved from https://cchs.gwu.edu/sites/cchs.gwu.edu/files/downloads/Covington_SpecialIssueBrief.pdf.

Canadian Security Intelligence Service (1994). *Economic security*. Retrieved from www.datapacrat.com/True/INTEL/CSIS/BACK6E.HTM.

Chertoff, M. (2011, 1 November) *Data sovereignty in the cloud: The issues for government, SafeGov*. Retrieved from www.safegov.org/2011/11/1/data-sovereignty-in-the-cloud-the-issues-for-government.

China, Germany working on cybersecurity deal (2016). *Vertretungen der Bundesrepublik Deutschland in der Volksrepublik China*, 17 March. Retrieved from www.china.diplo.de/Vertretung/china/de/__pr/2016/reden__bo/160311-caixin-pm.html?archive=3366876.

Cyber security strategy for Germany (2011). *Federal Ministry of the Interior*, 2011. Retrieved from www.bsi.bund.de/SharedDocs/Downloads/EN/BSI/Publications/CyberSecurity/Cyber_Security_Strategy_for_Germany.pdf?__blob=publicationFile.

Cyber security strategy of the European Union: An open, safe and secure cyberspace (2013). European Commission, Joint Communication to the European Parliament, the Council, the European Economic and Social Committee and the Committee of the Regions, JOIN(2013), 2 February.

Declaration on principles of international law concerning friendly relations and cooperation among states in accordance with the Charter of the United Nations, 1970. UN General Assembly, UN Doc. A/RES/2625(XXV).

Declaration on the inadmissibility of intervention in the domestic affairs of states and the protection of their independence and sovereignty, 1965. UN General Assembly, UN Doc. A/RES/2131(XX).

Definition of aggression, 1974. UN General Assembly, UN Doc. A/RES/29/3314.

Delerue, F. (2016). State-sponsored cyber operations and international law (Doctoral thesis, European University Institute).

Fact sheet: President Xi Jinping's state visit to the United States (2015). Office of the Press Secretary, The White House, September 2015. Retrieved from www.whitehouse.gov/the-press-office/2015/09/25/fact-sheet-president-xi-jinpings-state-visit-united-states.

Fidler, D. (2013a). Economic cyber espionage and international law: Controversies involving government acquisition of trade secrets through cyber technologies. American Society of International Law. *Insights*, *17*(10). Retrieved from www.asil.org/insights/volume/17/issue/10/economic-cyber-espionage-and-international-law-controversies-involving.

Fidler, D. P. (2013b, 11 February). *Why the WTO is not an appropriate venue for addressing economic cyber espionage*. Retrieved from https://armscontrollaw.com/2013/02/11/why-the-wto-is-not-an-appropriate-venue-for-addressing-economic-cyber-espionage/.

Froman, M. B. G. (2015). *US special 301 report*. https://ustr.gov/issue-areas/intellectual-property/special-301/2015-special-301-review.

The Global Risks Report (2015). 10th edn, World Economic Forum. Retrieved from www3.weforum.org/docs/WEF_Global_Risks_2015_Report15.pdf

The Global Risks Report (2016). 11th edn, World Economic Forum. Retrieved from www3.weforum.org/docs/Media/TheGlobalRisksReport2016.pdf.

Gorman, S. (2011, 21 December). China hackers hit US Chamber. *Wall Street Journal*. Retrieved from www.wsj.com/articles/SB10001424052970204058404577110541568535300.

G20 Leaders' Communiqué Antalya Summit (2015, 15–16 November). Retrieved from http://g20.org.tr/g20-leaders-commenced-the-antalya-summit/.

The IP Commission Report (2013). The Report of the Commission on the theft of American intellectual property. The National Bureau of Asian Research. Retrieved from www.ipcommission.org/report/ip_commission_report_052213.pdf.

Kuchler, H., & Sevastopulo, D. (2014, 10 June). Second China unit accused of cyber crime. *Financial Times*. Retrieved from www.ft.com/cms/s/0/3a1652ce-f027-11e3-9b4c-00144feabdc0.html#axzz4HrE7pol2.

Laney, D. (2001). *3D Data management: Controlling data volume, velocity, and variety. Application delivery strategies, meta group*. Retrieved from https://blogs.gartner.com/doug-laney/files/2012/01/ad949-3D-Data-Management-Controlling-Data-Volume-Velocity-and-Variety.pdf.

Lewis, J. A. (2010, March). *The cyber war has not begun*. Center for Strategic and International Studies. Retrieved from https://csis-prod.s3.amazonaws.com/s3fs-public/legacy_files/files/publication/100311_TheCyberWarHasNotBegun.pdf.

Mandiant Report (2013). *APT1: Exposing one of China's cyber espionage units*. Retrieved from www.fireeye.com/content/dam/fireeye-www/services/pdfs/mandiant-apt1-report.pdf.

McAfee Report (2013). *The economic impact of cybercrime and cyber espionage*. Retrieved from www.mcafee.com/us/resources/reports/rp-economic-impact-cybercrime.pdf) http://djilp.org/4721/critical-analysis-economic-espionage-and-international-law/.

Merkin, K. (2013, 26 December). *Critical analysis: Economic espionage and international law. The view from above*. Retrieved from http://djilp.org/4721/critical-analysis-economic-espionage-and-international-law/.

Nakashima, E. (2013, 11. March). US publicly calls on China to stop commercial cyber-espionage, theft of trade secrets. *Washington Post*. Retrieved from www.washingtonpost.com/world/national-security/us-publicly-calls-on-china-to-stop-commercial-cyber-espionage-theft-of-trade-secrets/2013/03/11/28b21d12-8a82-11e2-a051-6810d606108d_story.html.

Protection against the unlawful acquisition of undisclosed know-how and business information (trade secrets) (2013, 28 November). European Commission. Retrieved from http://europa.eu/rapid/press-release_MEMO-13-1061_en.htm.

Report of the Group of Governmental Experts on Developments in the Field of Information and Telecommunications in the Context of International Security (2013, 24 June). UN Doc. A/68/98.

Rogin, J. (2012, July). NSA Chief: Cybercrime constitutes the greatest transfer of wealth in history. *The Cable*. Retrieved from http://foreignpolicy.com/2012/07/09/nsa-chief-cybercrime-constitutes-the-greatest-transfer-of-wealth-in-history/.

Trans Pacific Partnership, US Trade Representative. Retrieved from https://medium.com/the-trans-pacific-partnership/intellectual-property-3479efdc7adf#.viwdwaq2c.

UK–China Joint Statement on building a global comprehensive strategic partnership for the 21st Century (2015, 22 October). Foreign and Commonwealth Office. Retrieved from www.gov.uk/government/news/uk-china-joint-statement-2015.

The United States Department of Justice (2009). *Former Boeing engineer convicted of economic espionage in theft of space shuttle secrets for China*. Retrieved from www.justice.gov/opa/pr/former-boeing-engineer-convicted-economic-espionage-theft-space-shuttle-secrets-china.

Judgements

Air Services agreement of 27 March 1946 between the United States of America and France,1978. R.I.A.A. Vol. XVIII.

Argentina – Certain measures on the protection of patents and test data, 2000. Document WT/DS196/1.

Argentina – Patent protection for pharmaceuticals and test data protection for agricultural chemicals, 1999. Document WT/DS171/1.

Case concerning military and paramilitary activities in and against Nicaragua (Nicaragua v. *United States of America)*, 1986. Merits, Judgment, ICJ Reports 14.
China – Measures affecting the protection and enforcement of intellectual property rights, 2009. World Trade Organisation, WT/DS362.
Corfu Channel Case, 1949. ICJ Reports 4.
Gabčíkovo–Nagymaros Project (Hungary v. *Slovakia)*, 1997. Judgment, ICJ Reports 7.
Naulilaa Incident arbitration (Portugal v. *Germany)*, 1928. 2 R.I.A.A. 1012.
Questions relating to the seizure and detention of certain documents and data (Timor-Leste v. *Australia)*, 2014. Provisional Order ICJ Rep 147.
SS Lotus Case (France v. *Turjey)*, 1927. PCIJ Report Series A No 10.

Index

For Product Safety Concerns and Information please contact our EU representative GPSR@taylorandfrancis.com
Taylor & Francis Verlag GmbH, Kaufingerstraße 24, 80331 München, Germany

www.ingramcontent.com/pod-product-compliance
Ingram Content Group UK Ltd.
Pitfield, Milton Keynes, MK11 3LW, UK
UKHW011002180926
475701UK00004B/527
* 9 7 8 0 3 6 7 2 2 7 5 6 2 *